Consciousness and Causality

Great Debates in Philosophy

Personal Identity
Sydney Shoemaker and Richard Swinburne

Consciousness and Causality
D. M. Armstrong and Norman Malcolm

Consciousness and Causality

A Debate on the Nature of Mind

D. M. Armstrong and Norman Malcolm

Basil Blackwell

© D. M. Armstrong and Norman Malcolm 1984

First published 1984
Basil Blackwell Publisher Limited
108 Cowley Road, Oxford OX4 1JF, England

All rights reserved. Except for the quotation of short passages for the purposes of criticism and review, no part of this publication may be reproduced, stored in a retrieval system, or transmitted, in any form or by any means, electronic, mechanical, photocopying, recording or otherwise, without the prior permission of the publisher.

Except in the United States of America, this book is sold subject to the condition that it shall not, by way of trade or otherwise, be lent, re-sold, hired out, or otherwise circulated without the publisher's prior consent in any form of binding or cover other than that in which it is published and without a similar condition including this condition being imposed on the subsequent purchaser.

British Library Cataloguing in Publication Data

Armstrong, D. M.
Consciousness and causality.—(Great debates in philosophy)
1. Mind and body 2. Intellect
I. Title II. Malcolm, Norman III. Series
128'.2 BD162

ISBN 0-631-13212-0
ISBN 0-631-13433-6 Pbk

Typeset at The Spartan Press Ltd, Lymington, Hants
Printed in Great Britain by Pitman Press Ltd, Bath

Contents

Great Debates in Philosophy

Since the time of Socrates, dialogue has been a powerful means of philosophical exploration and exposition. By presenting important current issues in philosophy in the form of a debate, this new series attempts to capture the flavour of philosophical argument and to convey the excitement generated by the interplay of ideas.

There will normally be more than two sides to any argument, and for any two 'opponents' there will be points of agreement as well as points of disagreement. The debate will not, therefore, necessarily cover every aspect of the chosen topic, nor will it present artificially polarized arguments. The aim is to provide, in a thought-provoking format, a series of clear, accessible and concise introductions to a variety of subjects, ranging from formal logic to contemporary ethical issues. The series will be of interest to scholars, students and general readers alike, since each book brings together two outstanding philosophers to throw light on a topic of current controversy.

The first essay states a particular position and the second essay counters it. The first author's rejoinder is again answered in the second author's reply. If the resulting book gives rise in its turn to further discussion, argument and debate among its readers it will have achieved its purpose.

Consciousness and Causality

NORMAN MALCOLM

Acknowledgements

I am grateful to the members of my seminars at King's College London, and at the University College of Swansea, in the academic year 1981–82, for their great help in our discussions of the topics of the present essay. I am especially indebted to Dan Rashid, İlham Dilman, Marina Barabas, and Dewi Phillips.

1 Consciousness

There is a grammatical difference between two uses of the word 'conscious'. In one use this word requires an object: one is said to be conscious *of* something, or to be conscious *that* so-and-so. One can be said to be conscious *of* a strange odour, of the stifling heat, of a friend's ironical smile; and one can also be said to be conscious *that* there is a strange odour in the house, that the room is stifling hot, that one's friend is smiling ironically. The expessions 'conscious of' and 'conscious that' are generally replaceable by 'aware of' and 'aware that'. Being conscious *of* something or *that* so-and-so, I shall call the 'transitive' use of the word 'conscious'; and I shall speak of 'transitive consciousness'.

There is another use of the word 'conscious' in which it does not take an object. If we think that a person who was knocked unconscious has regained consciousness, we can say, 'He is conscious', without needing to add an 'of' or a 'that'. This use, in which someone can be said to be conscious or unconscious *tout court*, I shall call the 'intransitive' use of the word 'conscious', and shall speak of 'intransitive consciousness'. It may be noted that when 'conscious' is used intransitively it cannot be replaced by 'aware': to be aware is always to be aware *of* or *that*.

Transitive consciousness

Transitive consciousness is consciousness of or that: consciousness in this sense requires an object. A long disputed question is: what is the relation of consciousness to the objects of consciousness? In a famous early paper entitled 'The Refutation of Idealism',[1] G. E. Moore attacked the doctrine, held by some idealists, that *esse* is *percipi*. Moore understood this doctrine to amount to the claim that *whatever is*, is *experienced*.[2] Apparently this meant that nothing that is an object of consciousness can exist *except as* an object of consciousness. In

[1] G. E. Moore, *Philosophical Studies*, Harcourt Brace & Co., 1922.
[2] Ibid., p. 7.

his discussion Moore concentrated on what he called 'sensations' or 'ideas'. He spoke of 'the sensation of blue' and 'the sensation of green'. He said that although these are different sensations they have something in common that he called 'consciousness'. That in respect to which the two sensations differed, he called a difference in the 'objects' of consciousness. Moore said: 'We have then in every sensation two distinct elements, one which I call consciousness, and another which I call the object of consciousness.'[3] Moore went on to say:

> The true analysis of a sensation or idea is as follows. The element that is common to them all, and which I have called 'consciousness', really *is* consciousness. A sensation is, in reality, a case of 'knowing' or 'being aware of' or 'experiencing' something. When we know that the sensation of blue exists, the fact we know is that there exists an awareness of blue.[4]

Moore said that awareness of or consciousness of something 'is involved equally in the analysis of *every* experience—from the merest sensation to the most developed perception or reflexion'.[5] He went on to make an assertion that pertains to my present topic. He said that this awareness or consciousness 'is and must be in all cases of such a nature that its object, when we are aware of it, is precisely what it would be, if we were not aware (of it)'.[6] This assertion of Moore's was a blow aimed at the doctrine of *esse* is *percipi*. This latter doctrine is certainly extravagant if it implies that chairs and mountains do not, or cannot, exist except when they are objects of consciousness. But did not Moore go too far in the other direction? Objects of consciousness include bodily sensations, such as pains and aches: for we can say, 'as I reached the summit I became conscious of a pain in my back'. Is a pain in one's back something of such a nature that, as Moore asserted, when one is aware of it it is precisely what it would be if one were not aware of it?

Armstrong on consciousness. Before treating this question we may note that David Armstrong has adopted a view of the

[3]G. E. Moore, p. 17. [4]Ibid., p. 24. [5]Ibid., p. 29. [6]Ibid.

nature of consciousness, which is in a certain way similar to the view that Moore took in his 'Refutation'. In *A Materialist Theory of the Mind*[7] Armstrong says:

> I suggest that consciousness is no more than *awareness* (perception) of inner mental states by the person whose states they are. . . . If this is so, then consciousness is simply a further mental state, a state 'directed' towards the original inner state.[8]

It would seem, from this passage, that Armstrong is using the term 'consciousness' in a narrower sense than Moore did. For Moore was using it in such a way that a person could be said to be conscious of a table or of the blue colour of a flower, neither of which are 'inner mental states'. But this difference in the employment of the term 'consciousness' is not relevant to the feature of Armstrong's view I want to consider. Armstrong says the following about the nature of awareness or consciousness:

> Let us consider the mechanical analogue of awareness of our own mental states: the scanning by a mechanism of its own internal states. It is clear that the operation of scanning and the situation scanned must be 'distinct existences'. . . . Now what reason is there to think that awareness of its own states in the case of, say, a spiritual substance, will differ in its logical structure from that of a self-scanning device in a mechanism? Why should the substitution of spiritual for material substance abolish the need for a distinction between object and subject? I must admit that I can see no way to prove that there must be such a parallelism, which is a *lacuna* in my argument. But it seems clear that the natural view to take is that pain and awareness of pain are 'distinct existences'. If so, a false awareness of pain is at least logically possible.[9]

Several things in this passage may be noted. First, the assumption that the 'logical structure' of human awareness or consciousness is the same as that of a self-scanning device in a

[7]D. M. Armstrong, *A Materialist Theory of the Mind*, Routledge & Kegan Paul, 1968. (Hereafter cited as *MTM*.)

[8]*MTM*, p. 94.

[9]*MTM*, pp. 106–7.

mechanism is surprising. I will come back in a moment to examine this analogy. Second, the distinction between 'spiritual substance' and 'material substance' is a red herring. It should not distract us: the concept of consciousness certainly applies to human beings, whether or not they are 'spiritual' or 'material'. Third, the phrase 'a false awareness of pain' is paradoxical. Probably it is supposed to mean the same as 'one's having a mistaken belief that one is in pain'. But the meaning of *this* expression is also uncertain, since the word 'belief' is not used like that in ordinary life—that is, we don't speak of anyone's either 'believing' or 'not believing' that he is in pain. Fourth, the meaning of 'distinct existences' is not transparent: nevertheless, there is no doubt that Armstrong, when he says that pain and awareness of pain are 'distinct existences', means *both* that a person could have 'a false awareness of pain', *and also* that a person could be in pain without being aware of it. The latter point is confirmed by the following remarks:

> A 'feeling of pain' is simply a *sensation* of pain. Now we can have a sensation of pain and be perfectly unaware of having it. So there can be a feeling of pain that we are unaware of feeling: unconscious pain.[10]

So, on Armstrong's view, there is as great a logical gap between pain and awareness of pain, as between dogs and dog-collars: just as one could have a dog without a dog-collar, or a dog-collar without a dog, so too it would be possible for a person to feel pain without being aware of feeling pain, and also to be aware of feeling pain without feeling pain.

I take it as fundamental that we cannot investigate the relation between the concept of pain and the concept of awareness unless we *remind ourselves* of how pain-locutions and awareness-locutions are actually employed in real life. When we do this it will be evident that there is not the radical separation between the concept of pain and the concept of awareness or consciousness, that Armstrong's theory requires. For example, suppose that an hour ago you com-

[10] *MTM*, p. 312.

plained of a severe pain in your shoulder. I might now ask you any of the following questions:

> Are you still in pain?
> Do you still have pain?
> Do you still feel pain?
> Are you still aware of pain?

Because of the different *wording* of these four sentences, a philosopher might be led to think that they must have different meanings. But in everyday life there would be no difference in how they were taken. Your answer of 'Yes' would have the same practical consequences in all four cases: likewise for an answer of 'No'. Of course you might have answered differently. You might have said, 'The pain is still there but it's not as bad as it was'; or, 'I'm conscious of pain but it doesn't trouble me as much as it did'; or, 'I have pain but it isn't as severe as previously'. These three replies, despite their different wording, would all be understood as coming to the same: no one would draw a different practical conclusion from one of them than from the other two.

Another possible reply would be this: 'I still have pain; but if I sit quietly, as I'm doing now, I am not aware of any pain; if I move about, then I feel it'. This reply shows that the sentence 'I have pain' does not always mean that the speaker is in pain at the moment of speaking. It can instead be used as an abbreviated conditional statement, coming to much the same as, 'If I move about or lift something, etc., then I am aware of (feel, am conscious of, have) pain'. The sentences, 'I feel pain', 'I have a feeling of pain', can also be given this same conditional use. Armstrong's assertion, 'there can be a feeling of pain that we are unaware of feeling', could be given a sensible interpretation in terms of this conditional use of 'I feel pain' or 'I have a feeling of pain'. One could say, 'I still have a feeling of pain in my shoulder, but only when I lift my arm like this: otherwise I am not aware of feeling pain'.

It is obvious, however, that this idiomatic way of speaking offers no support for Armstrong's philosophical thesis that pain and awareness of pain are 'distinct existences'. The meaning of this thesis is best revealed by Armstrong's opinion that the awareness of a sensation has the same 'logical structure' as a self-scanning device of a machine. Let us look into this comparison. Suppose that an automobile engine is equipped with a temperature gauge that will measure the heat of the engine. If the engine is overheating the device will activate an electric warning light on the dashboard, so that the driver may know of the danger. Do this self-scanning device and a person's awareness of a sensation, have the same 'logical structure'? In raising this question, Armstrong, I assume, wants us to compare the *concept* of such a device with the *concept* of the awareness of a sensation: or, to put it in another way, to compare what is logically possible, or conceivable, or what it makes sense to say in the one case as compared with the other.

In regard to the device that is supposed to indicate engine overheating, one possibility is that although the device had been manufactured it was never actually installed in any car, so that the frequent overheating of car engines was never indicated by the device. Another possibility is that although the device was usually installed in cars, in many or even most cars this was done incorrectly, so that in most cars no warning signal appears when the engine is overheating and also the signal often appears when the engine is not overheating. Another possibility is that even if the device was correctly installed, the materials and wiring of the gauge suffer from wear and age, so that after a few years overheating fails to be signalled or is falsely signalled. And so on.

In the analogy, cars correspond to people, the overheating of engines to sensations of pain, and the signalling or registering of overheating by the self-scanning device corresponds to one's consciousness of one's own sensations of pain. According to the analogy the following are logical possibilities: that all human beings might go through life having many sensations of pain but without ever being conscious of any of them; or that most people are so constituted that not only is it the case that whenever they have sensations of pain they are not conscious of them, but also that whenever they seem to themselves to be conscious of sensations of pain they are in fact having no sensations of pain; or that usually everyone starts out in life with the 'right' constitution, but after some years nearly everyone's consciousness of sensations of pain becomes defective and inaccurate.

In order to fully appreciate how bizarre are these alleged possibilities one should understand that what is implied is not just that these supposed possibilities might be realized in some science-fiction never-never land, but instead that they may be what is actually the case here and now with ourselves. The analogy implies, for example, that most or even all of us, including you and me, may be so constituted that we have frequent sensations of pain without ever being aware of them, and also that whenever we seem to ourselves to be aware of sensations of pain there may actually be no such sensations present. Clearly there is something radically wrong with the suggestion that this may be the situation with us right now. What is wrong with it?

It is not difficult to see that the supposed possibilities are so contrary to the way we speak and think of our sensations and of our awareness of them, that to the extent we took these 'possibilities' seriously then to the same extent we would realize that we had *no understanding* of the language of sensation — that we no longer knew how to use such expressions as 'a sensation of pain', 'a feeling of pain', 'aware of a feeling of pain', 'conscious of a sensation of pain', etc. But if we lost our grip on such expressions, then also we would fail to understand the supposed possibilities, since they are stated in terms of those very expressions! These 'possibilities' are thus seen to be self-destructive and in that sense incoherent. Yet these supposed possibilities are implied by Armstrong's thesis that pain and awareness of pain are 'distinct existences', and that one's consciousness of one's own sensations has the same 'logical structure' as a self-scanning device in a machine.

If the concept of consciousness did resemble the concept of a self-scanning device then, since there might be a machine that had not been equipped with such a device, or the device had been incorrectly installed or had become defective, so that the machine was not 'aware' of its states—so also there might be a human being who is suffering severe pain but is not aware of any pain! What would that mean? If the person was not aware of any pain at all, in the normal meaning of these words, then there ought to be an absence of any of the pain-behaviour that is natural to human beings: no contortions, grimaces, groans, outcries, etc.; and also nothing interfering with work or play; no seeking of help or comfort; no complaining of pain; and so on. Armstrong is asking us to conceive of the possibility of there being a person who gives every appearance of being healthy, vigorous, working efficiently, in a cheerful good humour, etc., but who is actually feeling intense pain! We cannot conceive of such a possibility, not from lack of imagination, but because the normal use of the expressions 'pain', 'feeling pain', 'sensation of pain', 'awareness of pain', etc., links all of them in a conceptual connection with the human behaviour, the reactions and actions, that are manifestations of pain—and does not link them merely causally and contingently with that behaviour, as Armstrong thinks.[11] But I will postpone until section 3 a study of Armstrong's causal theory of mind.

Dennett on the inconsistency of the concept of pain. In philosophy there has been a huge amount of controversy over the topic of pain. There are disputes over whether there can be unfelt pain, whether one can feel pain of which one is unconscious, whether one can mistakenly think one is in pain or make a mistake in locating the pain, whether a computer could feel pain, whether two people could have the same pain, whether pain is a 'private' sensation, and so on. The topic of pain is a good one for bringing to the surface a lot of philosophical confusion about the concept of sensation.

Recently it has been claimed that this turmoil is not so much due to the confusions and crude imagery that we philosophers bring to the topic, as to the fact that the ordinary use of the word 'pain' is inconsistent or incoherent. Daniel Dennett has made this claim. K. V. Wilkes says: 'Dennett has argued persuasively that the everyday concept of pain is internally inconsistent . . .'[12] Let us see just how persuasively Dennett has argued. In his book *Brainstorms* he says:

> The ordinary use of the word 'pain' exhibits incoherencies [*sic*] great and small. A textbook announces that nitrous oxide

[11] *MTM*, e.g. p. 312, p. 313.

[12] K. V. Wilkes, 'Functionalism, Psychology, and the Philosophy of Mind', *Philosophical Topics* 12, 1981, p. 165.

> renders one 'insensible to pain', a perfectly ordinary turn of phrase which elicits no 'deviancy' startle in the acutest ear, but it suggests that nitrous oxide doesn't prevent the occurrence of pain at all, but merely makes one insensible to it when it does occur (as one can be rendered insensible to the occurrence of flashing lights by a good blindfold). Yet the same book classifies nitrous oxide among analgesics, that is *preventers* of pain (one might say 'painkillers') and we do not bat an eye.[13]

Dennett thinks he has here presented evidence that the ordinary use of the word 'pain' is incoherent. The evidence is that most readers of the textbook do not perceive the difference in meaning between saying that nitrous oxide renders one 'insensible to pain' and saying that it 'prevents pain'. But why does Dennett believe there *is* a difference in meaning? Clearly because his attention is fixed on a form of words and not on how that form of words is *used*. He says that the phrase 'insensible to pain' *suggests* that the pain is still there. To *whom* would it suggest this? Perhaps to a philosopher puzzled about the concept of consciousness: but not to a medical student who wants to learn how to relieve patients of pain in surgery. Dennett's comparison ('as one can be rendered insensible to the occurrence of flashing lights by a good blindfold') reveals his assumption that the phrase 'insensible to . . .' does have, or should have, the same meaning in every context. But that is not how our language works. If in some contexts, but not in others, the expressions, 'renders one insensible to X' and 'prevents X', are given the same meaning, then that is how they are used! There is no basis here for a charge of incoherence.

Another alleged example of incoherence in the ordinary use of the word 'pain' is derived from the notion that different people have different 'pain-thresholds'. Dennett says:

> Some people, it is often claimed, can stand more pain than others: they have a *high* pain threshold. Suppose I am one of those with a *low* threshold, and undergo treatment (drugs, hypnosis, or whatever) supposed to change this. Afterwards I report it was a complete success. Here is what I say:
>
> (1) The treatment worked: the pain of having a tooth drilled is as intense as ever, only now I can stand it easily. Or I might say something different. I might say:

[13]Daniel C. Dennett, *Brainstorms*, Harvester Press, 1978, p. 221.

> (2) The treatment worked: having a tooth drilled no longer hurts as much: the pain is less severe.
> Can we distinguish these claims? Of course. They obviously mean very different things. Can I then know which claim is correct in my own case or in another's?[14]

There is indeed an incoherence here. It lies in Dennett's first imagined report. It would certainly be surprising for someone to say that the treatment 'worked', was 'completely successful', 'I can now stand the drilling easily'; but, in the same breath, to add that 'the pain of having a tooth drilled is as intense as ever'. This combination of remarks would be apparently contradictory. If in addition to these remarks there had been a marked diminution of pain-behaviour during drilling, then this first report would probably be taken as a jocular way of saying that the pain of drilling was less severe. In which case the two reports would come to the same.

In this connection it may be noted that both Dennett and Armstrong attach philosophical significance to a phenomenon that sometimes occurs when patients undergo prefrontal lobotomy. According to Dennett the patients say that they continue to be in pain, but they 'seem and claim not to *mind*' the pain.[15] Armstrong says:

> Intractable pain can sometimes be removed by severing connections between the prefrontal lobes and the rest of the brain. But patients on whom this operation has been performed sometimes give very curious reports. They say that the pain is still there, but it does not worry them any more. It seems as if they are saying that they have a pain which is giving them no pain![16]

This phenomenon is indeed significant. But in what way? If a patient said that the pain was as *intense as ever* but that he didn't *mind* it anymore, then we should not understand what he is saying. If the patient displays *none* of the normal behaviour of pain, but insists that 'the pain is as intense as ever', one would be baffled by his declaration. One would not understand what he is saying.

[14] Ibid., p. 222. [15] Ibid., p. 221.
[16] Armstrong, *MTM*, p. 313.

G. E. M. Anscombe imagines an example that is somewhat similar. She is referring to one's ability to say *where* one feels pain, and to the fact that what one says is normally accepted. She remarks, however, that one can imagine circumstances in which it would *not* be accepted. For example:

> If you say that your foot, not your hand, is very sore, but it is your hand you nurse, and you have no fear of or objection to an inconsiderate handling of your foot, and yet you point to your foot as the sore part: and so on. But here we should say that it is difficult to guess what you could mean.[17]

In the examples of lobotomy, and also in Anscombe's imaginary case, what the person says is not understood since it is *at odds* with other parts of his behaviour. Both cases illuminate the tie between the language of pain and the rest of one's behaviour, by showing that when these conflict, then, as Anscombe says, it is difficult to guess what the speaker means. Such cases, real and imaginary, do not reveal any inconsistency in the ordinary use of the word 'pain', but instead reveal that the use of this word is more complex than perhaps one had supposed.

To continue with Dennett's argument that the ordinary use of the word 'pain' is contradictory: he says that

> The grammatical grounds for the contradiction have already been noted: it is equally ordinary to speak of drugs that prevent pains or cause them to cease, and to speak of drugs that render one insensitive to the pains that may persist. . . . So ordinary usage provides support for the view that for pains, *esse est percipi*, and for the view that pains can occur unperceived. . . . What must be impeached is our ordinary concept of pain. A better concept is called for . . .[18]

Dennett brings this contention of his to bear on the much disputed question of whether it would be possible to make a computer or robot that feels pain. His answer is in the negative on the ground that, since the ordinary concept of pain is inconsistent, there can be no 'true theory' of pain that could be incorporated into the design of a computer:

[17] G. E. M. Anscombe, *Intention*, Blackwell, 1967, p. 14.
[18] Dennett, *Brainstorms*, p. 225.

> If, as I have claimed, the intuitions we would have to honor were we to honor them all do not form a consistent set, there can be no true theory of pain, and so no computer or robot could instantiate the true theory of pain, which it would have to do to feel real pain.[19]

One can only hope that there is a better account of why a computer that feels pain cannot be manufactured!

Let us linger a bit with the notion that the ordinary use of pain-language is inconsistent. If a dentist were drilling a tooth of mine he might pause to ask, 'Do you have any pain?' Or he might ask, 'Do you feel any pain?', or 'Are you aware of any pain?', or 'Are you conscious of any pain?', or 'Do you notice any pain?', or even 'Are you aware of any feeling of pain?', or just 'Does it hurt?' If one merely looks at these sentences, without considering how they are used, one might think that of course they mean different things. But as they are actually used in a dental office they all come to exactly the same. The answer 'No' to any of them would be the same answer to all of them. Likewise for the answer 'Yes'. If I answered 'Yes' the dentist might inject novocaine; if 'No' he would just carry on.

There could be other answers. To any of the questions I might answer, 'A little bit'. Would this mean a little bit of *pain*, or a little bit of *consciousness* of possibly a lot of pain? A comical question. The answer 'A little bit' would mean the same throughout this range of questions, namely, a little bit of pain.

Does this mean that according to ordinary pain-language, the *esse* of pain is *percipi*? I would not say so because, first, this would make it appear that ordinary pain-language embodies a philosophical theory, which it doesn't; and second, one does not normally speak of 'perceiving' one's own pain or one's feeling of pain. (A third reason will be mentioned shortly.) But this brief reminder of how a dentist and his patient speak to one another, does show that Moore was wrong in holding of *every* 'object of consciousness' or 'object of awareness', that 'this awareness is and must be in all cases of such a nature that its object, when we are aware of it, is precisely what it would be, if we were not aware (of it)';[20] and it also shows that Armstrong is

[19] Ibid., p. 228. [20] Moore, *Studies*, p. 29.

wrong in holding that 'pain and awareness of pain are "distinct existences"'.[21]

Being distracted. Dennett might have offered more plausible-seeming evidence of incoherence in ordinary pain-language. I have in mind the fact that one's attention can be distracted from pain. Suppose that during a long walk I began to feel an aching in my legs. But then my companion and I started up a lively conversation during which I ceased to be aware of the aching; yet when the interesting talk ended I became conscious again of the aching. Could I report that the aching *ceased* during the conversation? No: for I did not notice that it had ceased. I did not think to myself, 'The aching has stopped!' That would be a possible case, but not the one I am describing. Could I report that the aching *continued* throughout the conversation? No: for I was not aware of any aching during that time. I *could* say, 'The conversation distracted my attention from the aching'; but this common way of speaking does not imply *either* that the aching continued *or* that it stopped. This example provides a third reason for rejecting the idea that, according to ordinary pain-language, the *esse* of pain is *percipi*. For this doctrine ought to imply that when I was no longer *aware* of any aching then the aching must have stopped. But in the case described I do not *want* to say that it stopped: all I want to say is that for a time I was not aware of any aching.

One might object that it *must* be either that the aching continued or that it stopped. But this 'must' comes from a false analogy: from assimilating the aching to a physical process (such as the flashing of a light), and from assimilating the phenomenon of one's attention being distracted from the aching to the phenomenon of one's attention being distracted from a flashing light. It makes sense to say that the flashing continued while I was distracted, and also to say that it stopped. One could investigate the matter. But we have no understanding of what it would be to investigate whether the aching continued or whether it stopped during the time that I was in the conversation.

In a C. S. Forester novel, Hornblower, the great sea captain, is on the deck of his ship during a severe storm. This sentence

[21] Armstrong, *MTM*, p. 107.

occurs: 'Hornblower found the keen wind so delicious that he was unconscious of the pain the hailstones caused him'. Does this mean that the hailstones caused him pain of which he was unconscious, or that they didn't cause him any pain? What if Forester had written instead: 'Hornblower found the keen wind so delicious that he was not aware of any pain from the blows of the hailstones'; or this: 'Hornblower found the keen wind so delicious that the blows of the hailstones did not cause him any pain'? In respect to our understanding of the situation, wouldn't these three sentences come to the same?

The phenomenon of having one's attention distracted or diverted from a bodily sensation is of philosophical interest. It shows that there is more than one concept of transitive consciousness (being conscious *of* something). There is a concept of consciousness that provides for the possibility of discovering whether or not an object of one's consciousness (e.g. the flashing of a light) did or did not exist or occur during a period when one was no longer conscious of it. There is another concept of consciousness (being conscious of a bodily sensation) which does not provide for such a discovery. This difference in the use of the expression 'conscious of . . .' shows that the question of the relationship between consciousness and the objects of consciousness, does not have a *single* answer, such as Moore and Armstrong have tried to give it. Nor is this double use of the phrase 'conscious of . . .' any evidence of an incoherence in the ordinary use of the phrase.

Indeterminacy. I want to consider again the notion of different 'pain-thresholds', not because it provides any support for Dennett's idea that the ordinary use of the word 'pain' is inconsistent, but because it leads into a topic of genuine significance, namely, what could be called an 'indeterminacy' in our attributions of sensations and feelings to people.

Suppose a woman has received a badly lacerated arm in an accident. She remains calm, does not make a fuss, and insists that the other injured people should be looked after first. Does she have less pain than would be normal with that sort of injury, or does she have a lot of pain but is bravely playing it down? Often we might not know which of these things to think. Perhaps not always: if she was someone whom we knew well, and knew that with ordinary minor injuries (hard knocks, sprains, slight burns) she complained as much and exhibited as much pain-behaviour as most people would, then we might be inclined to think, not that she has a 'high pain-threshold', but rather that on this occasion, perhaps from concern for the other injured people, she was suppressing the natural manifestations of severe pain; or perhaps that the excitement of the accident and the sight of the injuries of others, had distracted her attention from her own pain.

It could, however, also be the case that there was no ground for thinking like that, yet some people would think she was in great pain and was being remarkably brave, others would think that despite the severity of the injury her pain was not great, and still others would be uncertain what to think. Even if she herself said, 'It's not so bad', this might not remove the uncertainty and disagreement. Also, there is the other kind of case, where a person who is slightly injured exhibits signs of severe pain: some observers think the pain is very acute, others think the behaviour is exaggerated, and still others are uncertain what to think. Such disagreement and uncertainty occur not merely in regard to signs of physical pain, but also in regard to expressions of emotions and attitudes, such as grief, remorse, friendly feeling.

Wittgenstein addresses himself to this topic in several passages in *Zettel*. One thing he calls attention to is the fact that most people would say that one feels nothing under general anaesthetic; yet some say, 'It *could* be that one feels something but forgets it completely'.[22] This is an interesting example of disagreement: everyone has the same evidence to go on; yet some have this doubt and some don't.

Wittgenstein mentions another example that is relevant to the idea of there being different pain-thresholds. He says:

> We often use the phrase 'I don't know' in a queer way; when for example we say that we don't know whether this man really feels more than that other, or merely gives stronger expression to his feeling. It is in that case not clear what sort of investigation could settle the question.[23]

[22]Ludwig Wittgenstein, *Zettel*, edited by G. E. M. Anscombe and G. H. von Wright, translated by G. E. M. Anscombe, Blackwell, 1967, §403.
[23]Ibid., §553.

In our day there is an inclination to suppose that this question could be decided if we knew more about what goes on in the nervous system. But if we did know more about what goes on *inside* people, in this literal sense, might there not still be the same uncertainty and disagreement that arises when *behaviour* is observed? Wittgenstein makes this comment:

> Imagine that people could observe the functioning of the nervous system in others. In that case they would have a sure way of distinguishing genuine and simulated feeling.—Or might they here again doubt whether the other feels anything when these signs are present?—At any rate it could readily be imagined that what they see there determines their attitude without their having any qualms.
> And now this can be transferred to outer behaviour.
> This observation fully determines their attitude to others and a doubt does not occur.[24]

It is a fact that people sometimes disagree, or are uncertain, as to whether someone's behaviour and words are genuine expressions of pain (or grief or friendly feeling), or whether these expressions are exaggerated or partly pretence. Sometimes the doubt is not idle, but is reflected in coldness or lack of sympathy. But others may have no doubt at all, and cannot even understand how there could be any doubt. These differences in taking, assessing, someone's demeanour and words are sometimes felt to be deeply troubling. Some people seem to be generally more suspicious, less trusting, than others. In a particular case one who trusts may not be able to convince one who doubts.

This fact helps to foster the idea that to discover what a person really feels would require the observation of what is *inside* him. But if this idea were to be taken literally and seriously, and if neural processes could be observed, might not the same doubt and disagreement occur that are now provoked by outer behaviour? It should not be taken for granted that observations of neural processes would necessarily put an end to uncertainty and disagreement. Might not some observers 'doubt whether the other feels anything when these signs are

[24]Ibid., §557.

present'? On the other hand, it could be imagined that those observations *did* result in complete agreement, did 'determine their attitude without their having any qualms'. But it could *equally well* be imagined that observations of a person's behaviour and circumstances, such as we have now, did *not* provoke any disagreement or uncertainty. All well-informed observers would be in agreement: 'This observation fully determines their attitude to others and a doubt does not occur.'

Wittgenstein says: 'Might not the attitude, the behaviour, of trusting, be quite universal among a group of people? So that a doubt about expressions of feeling is quite foreign to them?'[25]

It might be thought that the disagreement that exists is to be explained by the absence of clear-cut *rules* for making attributions of feelings to others. It is true there are no clear rules. But this is not an explanation of the disagreement. Instead it is the other way round. There can be clear, precise rules only when there is virtually universal agreement. The certainty of arithmetic, of judgements of measurement, of the relationships between colours, depend on this. The firmness and clarity of rules is a reflection of overwhelming agreement. That kind of agreement is not present in regard to people's sensations and feelings.

In the *Philosophical Investigations*[26] a striking contrast is presented:

> There is such a thing as colour-blindness and there are ways of establishing it. There is in general complete agreement in the judgements of colours made by those who have been found normal. This characterizes the concept of a judgement of colour.
>
> There is in general no such agreement over the question of whether an expression of feeling is genuine or not. I am sure, *sure*, that he is not pretending; but a third person is not. Can I always convince him? And if not is there some mistake in his reasoning or observations? 'You don't understand anything!' —we say this when someone doubts what we recognize as clearly genuine—but we cannot prove anything.[27]

[25] Ibid., §566.

[26] Ludwig Wittgenstein, *Philosophical Investigations*, edited by G. E. M. Anscombe and R. Rhees, translated by G. E. M. Anscombe, Blackwell, 1967. (Hereafter referred to as *PI*.)

[27] *PI*, p. 227.

There is an indeterminacy in judgements of feeling that is not present in arithmetical calculations, or when it is being judged whether a certain colour is between red and yellow. The indeterminacy is a consequence of lack of general agreement. There might be conjectures about the *causes* of this lack of agreement: but it is of greater importance for philosophy to realize that it exists! We could imagine a society in which disagreement and uncertainty about expressions of feeling did not occur at all or hardly ever. Their judgements of the genuineness of an expression of feeling would be just as determinate as our judgements of colour. Those people might have little or no tendency to think of the sensations, emotions, moods, thoughts, of other persons as being hidden and inaccessible.

I am not saying that with us no considerations may be adduced to try to sway another person to accept one's own response to an expression of feeling. If A and B had been fierce professional rivals for many years, and if A had often made derogatory comments about B, but when B died A showed signs of grief, then another person, C, might think that A's exhibition of grief was insincere or exaggerated. But someone might point out that A and B had been close friends when they were younger, and had been of help to one another in many trying situations. This would be an attempt to establish that B's death is for A an occasion for genuine grief. This information might influence C's attitude—but also it might not. As Wittgenstein says: we cannot *prove* anything.

There is some temptation to think that when there is disagreement in assessments of expressions of feeling, this is because people differ in the sensitivity of their perception of others. It is of course true that some people are more perceptive than others. What I mean, however, by 'indeterminacy' is that there are frequent examples of expressions of feeling, where equally sensitive and informed observers differ in their assessment of the sincerity of the expressions, and where no one can see any way of resolving the disagreement. This means that here there is no definite right or wrong in the matter. One person in such a dispute may say 'You are too gullible', and the other may say 'You are too insensitive', but this is only scolding. Each of the disputing persons may continue to say 'I'm right and you are wrong', but the sense of such remarks is different here from what it is in contexts of arithmetical calculation or measurments of length.

Of course there are countless situations where the behaviour and circumstances of a person are such that every observer will agree either that the person is pretending or that the expression of feeling is genuine. If there were not *that* much agreement there would be no shared concepts of pretence or of genuineness, and therefore no possibility of agreement or disagreement in their application.

Still, there are frequent cases of *insoluble* disagreement or uncertainty. I have no doubt that one important source of the constant tendency of philosophers to speak of all feelings, experiences, mental states, etc., as being 'inner', and to regard them as 'hidden', is just this uncertainty and disagreement that often characterize our responses to expressions of feeling. If this is so then the idea that *all* mental phenomena are inner and hidden is doubly erroneous: for, first, in a multitude of cases the circumstances and the human expressions of sensation or emotion are so unambiguous as to leave no doubt; and, second, in those cases where there is real indeterminacy it is a muddle to think there is something 'hidden', since that confuses indeterminacy with lack of knowledge.

Being aware of being conscious. I wish to consider a perplexing question related to the concept of transitive consciousness, namely, whether when you are conscious of something it is possible for you to discern, or perceive, or be aware of *your consciousness* of that something, i.e. can your consciousness of an object be an object of your consciousness? G. E. Moore had something to say about this question in his 'Refutation of Idealism'. Previously I mentioned his view that in every sensation there are two distinct elements; one is consciousness, the other is the object of that consciousness. In referring to his example of 'the sensation of blue', he said that one element of this sensation is the object *blue*, and the other element is the awareness or consciousness of that object. Moore went on to say:

> Though philosophers have recognized that *something* distinct is meant by consciousness, they have never yet had a clear

> conception of *what* that something is. They have not been able to hold *it* and *blue* before their minds and to compare them, in the same way in which they can compare *blue* and *green*. And this for the reason . . . that the moment we try to fix our attention upon consciousness and to see what, distinctly, it is, it seems to vanish: it seems as if we had before us a mere emptiness. When we try to introspect the sensation of blue, all we can see is the blue: the other element is as if it were diaphanous. Yet it *can* be distinguished if we look attentively enough, and if we know that there is something to look for. My main object in this paragraph has been to try to make the reader see it . . .[28]

Moore speaks of the extreme difficulty of discerning by introspection the element of consciousness. Is this difficult because the consciousness vanishes when one tries to focus on it (as a slippery thing may elude us when we try to take hold of it)? Or is the difficulty of a different sort, namely, that it makes no sense to speak of discerning, perceiving, becoming aware of, one's own consciousness of something?

Moore says not only that the element of consciousness *can* be distinguished from the object of consciousness, but also that he is trying to make the reader *see* it. Suppose that one of Moore's readers were to report to him as follows: 'It was difficult, but I finally did succeed in perceiving the element of consciousness in my sensation of blue colour'. How could anyone know that this reader had perceived what *Moore* called 'the element of consciousness'? How could we be satisfied that the two of them were referring to the *same* thing?

It is a puzzling idea that one could perceive, or fail to perceive, one's consciousness of something. On a walk in the country my companion might say to me, 'Do you see the blue colour of those distant woods?' Suppose my answer is 'Yes' and that then my companion further asks, 'Do you perceive your consciousness of the blue colour?' What a strange question! What could I answer? If I were to reply, 'Yes, I do perceive my consciousness of the blue colour', would that give any more information than did my previous simple reply, 'Yes, I see the blue colour'?

That it should be *questioned* whether one can note, perceive, be aware of, one's own perception or consciousness of some-

[28] Moore, *Philosophical Studies*, p. 25.

thing, may seem surprising. For isn't it a fact of experience that I see the blue colour; and don't I report this fact when I say, 'I see the blue colour'? My report seems to refer to two elements: the blue colour *and* my seeing it. How can I make such a report unless I perceive both elements? Moore's belief that if he was very attentive he could distinguish within the sensation of blue, the two elements, *blue* and *consciousness*, may have been derived from the following reasoning: 'My sensation of blue must contain both blue and consciousness; and so if I focus my attention carefully enough I *must* be able to perceive not only the blue colour but also my consciousness of it.'

This reasoning, however, is based on what Wittgenstein calls 'a grammatical illusion'. The illusion is to think that since the sentence 'I am conscious of the blue colour' can be a true report of my experience, then it must be possible for me to observe both the consciousness and the colour. Whereas in fact the report derives solely from an observation of the colour, not of the consciousness. But it is misleading for me to put the matter in that way—as if it were clear what would be meant by 'my observing my consciousness of the colour'.

John Locke made some striking comments on this topic. He says: 'Consciousness is the perception of what passes in a man's own mind. Can another man perceive that I am conscious of anything, when I perceive it not myself?'[29] Locke thinks this rhetorical question is obviously to be answered in the negative. He declares that it is 'impossible for any one to perceive without *perceiving* that he does perceive. When we see, hear, smell, taste, feel, meditate, or will anything, we know that we do so.'[30]

Locke is speaking of consciousness in the transitive sense — being conscious *of* something. One might criticize the generality of Locke's claim, noting for example that a jealous thought can occur to a man without his being aware of it *as a jealous thought*: his recognition of this character of the thought might not occur at all, or might come only after much reflection. But my concern here is with something more basic: namely, to what extent is it meaningful to speak of someone's perceiving or

[29]John Locke, *An Essay Concerning the Human Understanding*, edited by A. C. Fraser, two volumes, Oxford University Press 1894, vol. I, book II, ch. 1, sec. 19.

[30]Ibid., book II, ch. 27, sec. 11.

observing or being aware of his own perceptions, thoughts, feelings?

Let us start with an example from sight. Suppose there is on a table before me a metal ball, the motion of which is controlled by electromagnetic forces. It is an experiment and my job is to report, from moment to moment, whether the ball is in motion or at rest: so I call out 'It's moving'; 'Now it's stopped', etc. In order to make these reports I must see and observe the ball. But do I observe my seeing? *Could* I do this? If I don't observe my seeing, do I perhaps observe *that* I see the ball? In Locke's terms, is it impossible for me to perceive the ball unless I perceive that I perceive it? These are strange questions.

Another person could observe the ball, *and* could also observe that I see the ball and its movements. How would he do the latter? Well, he would observe that my head and eyes follow the motion of the ball; and he would hear my verbal reports (which he knows to be correct from his own observations of the ball). In short, he observes my behaviour, including what I say. I can observe the movements of the ball, or fail to observe them. But do we understand the supposition that I might observe, or not observe (perceive, be aware of) my own observings of the ball?

There is a great inclination to think that an outsider's observation, *via* behaviour, of a person's seeing, awareness, consciousness of something, is at best *second-hand*, and that the person's own inner perception of his seeing or awareness, is *first-hand* and *direct*. May it be that this is a complete mistake; that the *only* observation or perception there can be, of someone's seeing or awareness, is observation *via* behaviour? If this were right, then the answer to Locke's question, 'Can another man perceive that I am conscious of anything, when I perceive it not myself?' would be: another man can perceive that I am conscious of a particular object; and whether *I* might or might not perceive that I am conscious of that object, is not an intelligible question.

Inner perception of mental phenomena. Franz Brentano undertook to give a general characterization of all 'mental phenomena'. In his *Psychologie vom Empirischen Standpunkt*[31] he says

[31]Franz Brentano, *Psychologie vom Empirischen Standpunkt*, edited by O. Kraus, two volumes, Felix Meiner, 1955, vol. I, book II, ch. 1, sec. 5.

that each 'mental phenomenon' is characterized by 'a direction upon an object'; each one 'contains something as object within itself'. For example, 'in imagination something is imagined, in judgement something is affirmed or denied, in love loved, in hate hated, in desire desired, etc.' Brentano says mental phenomena (*psychische Phänomene*) can be defined as those phenomena 'which contain an object intentionally within themselves'. This reference to objects, says Brentano, is indicated by language: 'We say that a person rejoices in or about something, that a person sorrows or grieves about something.'

One may question whether Brentano's characterization is universally true. It seems that people are sometimes anxious without their anxiety having any object, or at least any object known to them. Sometimes the same is true of depression. And one may start the day feeling cheerful, or feeling gloomy, without being cheerful or gloomy about anything in particular.

This particular view of Brentano's is, however, sufficiently familiar, and I do not wish to comment further on it. Brentano has another view that has not been so well noted. He holds that a further general characteristic of all mental phenomena is that 'they are perceived only in inner consciousness' (*nur in innerem Bewusstsein wahrgenommen werden*).[32] Brentano calls the perception of mental phenomena 'inner perception' (*innere Wahrnehmung*); and he adds that so-called 'outer perception' is strictly speaking not perception at all, so that mental phenomena are 'the only ones of which perception [*Wahrnehmung*] in the proper sense of the word is possible'.[33] Finally, Brentano says that 'no mental phenomenon is perceived by more than a single individual'.[34]

Under 'mental phenomena' Brentano includes recalling, expecting, judging, conviction, doubt, desire, intention, anger, joy, sorrow, etc. What Brentano is holding is that whenever a person desires something, or expects something, etc., that person has an inner perception of his own desire or of his own expectation. And in addition Brentano holds that that person is the only one who can perceive that desire or that expectation.

[32]Ibid., sec. 6. [33]Ibid. [34]Ibid.

Brentano's views are bewildering. Don't I often perceive another person's expectation? And when I expect something do I perceive *my* expectation? A man makes preparations for a walk in the fields: his dog responds with a display of excited expectation, which the man perceives. There is an inclination to think that the man does not actually perceive the dog's expectation, but only the dog's expectant behaviour. Are we to think that the dog perceives its own expectation? Are we able to distinguish two mental states of the dog: first, its expectation of being taken for a walk; second, its perception or awareness of its expectation?

It might be said that we are unable to employ this distinction with dogs, but we certainly can with people. Let us consider this. Suppose that I am travelling by train to a town. A friend wrote to me that he would meet me at the station. If before or during the trip someone were to ask me, 'Do you expect to be met?', I would answer 'Yes'. Now does this imply that I perceived, or was aware of, my expectation? A strange question! Certainly, I will not say, 'No, I did not perceive it'. Do I have to affirm, therefore, that I did perceive it? But what would it *mean* to say this? Would it come to anything other than what was already said in the description of the case, namely, that if asked whether I expected to be met I would have said 'Yes'? If that is what it comes to, then the question of whether this conditional fact implies that I perceived my expectation, would be the vacuous question of whether this conditional fact implies itself.

According to Brentano every 'mental phenomenon' (joy, doubt, fear, sorrow, intention, etc.) is directed upon an object; and further, this mental direction upon an object is itself the object of an inner consciousness or inner perception. If a person is joyous his joy is directed on a particular object, and also the person has an inner consciousness of his joy. My concern is with this latter claim. *When* does a person have an inner perception of his joy—and for *how long*? He has it whenever and only when he is joyous, according to Brentano. So if someone had heard of Brentano's doctrine of inner perception, and wanted to learn what it is and when it occurs, he would learn that a person has an inner perception of sorrow when and only when he is sorrowing, and an inner perception of fear when and only when he is afraid, etc. He would have learned that Brentano's doctrine provides nothing more than a movement in language from 'He is afraid' to 'He has an inner perception of fear'.

This criticism is, however, too harsh. There is more here than the substitution of a distorted way of speaking for a familiar one. The 'more' is a mental picture, which is hard to describe but seems to be something like this: one's intent look is always fixed upon an inner space into which nothing can enter without being perceived. Now why do we, not just Locke and Brentano, but all of us, have a strong inclination towards this picture? This is a hard question, requiring one to dig into one's philosophical unconscious.

My hunch is that part of the motivation for the idea that there is an inner perception of every mental phenomenon is the following: when a person is joyful, sorrowing, frightened, etc., there is often some expression of this in his behaviour or speech. This is not necessarily or always so, but it is frequently so: if it were not we would not have common concepts of joy, sorrow, etc. Suppose now that a man has just received some great good news. One possible reaction of his would be to exclaim to the person who delivered the news, 'I am overjoyed!' Another possible reaction would be to break forth in a spontaneous dance of joy. The philosophical thinking I have in mind is this—unless the man had an inner perception of his joy he would not *realize* that either the exclamation or the dancing was *from joy*! He would not know that he was dancing from joy, or that it was from joy that he exclaimed 'I am overjoyed'. But, the reasoning continues, it is absurd to suppose that a person should not realize that he is dancing from joy, etc. Therefore he must perceive his own joy.

In criticism of this thinking I will focus on the assumption that if a man is dancing from joy he must *realize* that he is dancing *from joy*. That a person might dance from joy is straightforward enough: but there is something problematic about the idea that he might realize, or not realize, that this is what he is doing. When in actual language would one use the words, 'He realized he was dancing from joy'? Let us describe a case in which this could be said. Suppose there is a man who has been continuously depressed for a couple of years. He never has an emotional reaction to any news, good or bad. He is indifferent, untouched, withdrawn. He says, 'What difference does it make?' His family and friends are concerned about him: they have frequently called his attention to his indifference and his apathy. He himself is aware that nothing moves him. But on *this* occasion, when he hears the good news, he leaps up and starts dancing around the room. He realizes, to his astonishment, that he is dancing *from joy*!

This example illustrates a natural use for the words, 'He realized he was dancing from joy'. It is a form of words that makes sense—but only in special circumstances. Not in just *any* situation where a person danced from joy (or slammed a door in anger, etc.) would it be intelligible to say, 'He realized he was dancing from joy'. If this is right, the philosophical reasoning which argues that a joyous person must always perceive his joy, because if he didn't then he would not always realize that he is speaking or acting from joy, is undermined.

Notice that I am not making the (dogmatic) assertion that it is always senseless to say, 'He realized that he was acting from joy, fear, anger, etc.' Instead I am asking, in what sort of circumstances could one say such a thing? In reflecting on this request, and in describing cases, one sees that the natural use of this form of words is limited. If Brentano or any other philosopher asserts that whenever a person acts out of fear, anger, gratitude, etc., 'the person realizes it', we see that this is not how these words are employed. In seeing that the claim is false in respect to our ordinary use of language, we also see that *we do not understand* what that use of language is in which it is supposed to be always true.

Let us consider another example. According to Brentano, if one dislikes another person one has an 'inner perception' of this dislike. Instead of denying that this ever makes sense, let us ask: in what circumstances would it make sense? Suppose that for a number of years Mr A had regarded his professional colleague, Mr B, with esteem and affection. Recently, however, A has begun to notice various mannerisms of B that irritate him, e.g. his long-winded stories, the overly hearty way he laughs at his own jokes, even the way he chews his food. A begins to reflect quite a lot on these annoying habits of B, and in the process it gradually dawns on him that he really *dislikes* B and is not fond of him at all. Sometime later A writes to a friend about this development, saying: 'As my inner perception of my feelings about B grew in me, I resolved to avoid him in the future'.

This employment by A of the phrase 'inner perception' is appropriate in the particular context of A's earlier attitude towards B followed by his gradually coming to view B in a new light. Of course this example does not show that *whenever* one person consciously dislikes another person it is right to say that he has an 'inner perception' of his dislike. Quite the contrary! In most of the situations of life, where a person dislikes something, or is angry, amused, taken by surprise, etc., we should not understand him if he spoke of his 'inner perception' of those attitudes and reactions. If I have been waiting in line for hours to obtain seats for an opera, and if a late arrival tries to slip into the line ahead of me, whereupon I angrily rebuke him, it would be hilarious nonsense if afterward I said that I had an 'inner perception' of my anger on that occasion. The Locke-Brentano thesis that *all* mental phenomena are perceived in 'inner consciousness' or 'inner perception' is false—for it contradicts the normal employment of these expressions.

But if not 'inner perception' then what about plain 'perception'? Locke asserted that it is impossible for a person to smell, see or hear something without 'perceiving' that he smells it, etc. A person may be thinking of how to repair a broken window, or he may see the changing colours of the ocean, or smell the odour of cooking cabbage. But to speak of his perceiving, or not perceiving, his thinking, seeing, smelling, would not be intelligible straight off. As in our previous examples, we can think of special contexts in which such things could be said. Suppose a friend had suffered a concussion, the only lasting consequence of which was that he had lost his sense of smell. This incapacity lasted for several years. One day he exclaims with excitement: 'I perceive that I am beginning to smell the odour of cabbage!' This unusual mode of expression would be justified by the dramatic nature of the situation. In respect of its informative character this exclamation would come to the same as 'I am beginning to smell the odour of cabbage!' The phrase, 'I perceive that' was a rhetorical flourish, natural enough in the circumstances. It did not mean that in addition to the smelling there was a perception of the smelling. Usually the first-person sentence, 'I perceive that I smell it', would have no clear sense, over and above the sense of 'I smell it'. *I* can perceive that *you* smell the cabbage: my perception would be based on observation of your behaviour and what you say. In the normal case *you* cannot be said to 'perceive' that you smell the cabbage: you just smell it.

Intransitive consciousness

If a man has been felled by a blow and is lying there inert, we might be uncertain whether he is conscious or unconscious. We might ascertain that he is alive (his heart is beating) and unconscious. Would it be possible to ascertain that he is dead and unconscious? No. This is not how the term 'unconscious' is used. We apply the words 'conscious', 'unconscious', only to those whom we think may be alive. When we know that a man is dead, the question of whether he is conscious or unconscious is eliminated. It might be thought that this is because 'He is unconscious' *follows from* 'He is dead': if we think he is dead we conclude he is unconscious, and so the *question* of whether he is or isn't conscious is eliminated. This is wrong. No one who has learned that a friend is dead would go on to speak of him as 'unconscious' except as a euphemism. A speaker at a memorial service for those who died in battle might refer to them as 'the unconscious dead'. Although 'unconscious' may be given this ceremonial use, it is not the more familiar use of the term in which if we think that a man is unconscious we may try to revive him, or hope that he will *regain* consciousness. Once we are sure he is dead we abandon such hope. 'He is dead' eliminates *both* 'He is conscious' and 'He is unconscious'. To put the point in another way: only if we believe that a person who has been struck down in an accident is or may be alive, can we think of him as being either conscious or unconscious.

A puzzling question is this: if a man is conscious does it follow that he is conscious *of* something? I think not. A person who has been knocked unconscious could show signs of returning consciousness (moving his arms, opening his eyes, sitting up, muttering) without its being apparent that he is conscious *of* any sights, sounds, sensations, or of anything at all. Being conscious *of* something entails being conscious, but not vice versa.

There are degrees of consciousness. A person who has just regained consciousness may be dazed, groggy, 'half-conscious', i.e. not fully conscious, not alert, not able to do things he could normally do. There are resemblances between this condition and that of being half-asleep, or drunk, or in a sleep-walker's trance. There is also another sense for the notion of 'degrees' of consciousness. Among people there are of course differences in discernment, intensity of reflection, originality of conception, creative powers. A person who is outstanding in these ways may be said to have 'a high degree of consciousness'.

The human pardigm. Not only of people but also of some other creatures (e.g. dogs, horses) we can say 'alive but unconscious'. But what of insects or worms? A swatted fly is still wriggling or crawling: it is still alive but is it conscious or unconscious? That is a queer question to ask about a fly. Why is this? Wittgenstein remarks that 'Only of a living human being and what resembles (behaves like) one, can one say: . . . it is conscious or unconscious.'[35] This is a rich saying. It means that living human beings are the paradigms of what can be said to be conscious or unconscious. If we wonder whether some non-human creature can be said to be conscious or unconscious we are implicitly inviting a comparison with the human body, the flesh and blood, the postures, gestures, and especially the expressive human face and eyes. It is by comparison with the human face that a dog's face is called a 'face' (whereas a worm has no face, and one feels doubtful about saying that a fly has one). Because of its resemblance to the human face, one can see curiosity, alertness, fright, in a dog's face; and much of a dog's behaviour is so humanlike that we don't hesitate to say that a dog expects, recognizes, remembers, etc.

With a fly there is not enough resemblance to provide more than a tiny basis for psychological predicates. It will react to a movement in its vicinity by flying away. Can we say that it was 'frightened' by the movement? Actually we do say this: but all we mean is that the movement caused it to fly away. A fly cannot display the behaviour of fright that a dog can, nor the behaviour of recognition, anticipation, anger, etc. It will

[35]Wittgenstein, *PI*, §281.

struggle to free itself from fly-paper: but is it terrified? Is it in pain? We don't know how to apply these terms to a fly. It isn't even clear whether we should say that the fly 'perceived' the movement that caused it to fly away. We cannot confidently attribute any form of consciousness to a fly. There is some temptation to think that this is because we don't know enough: some day science may find out whether flies possess consciousness. This isn't so. The fly is so distant from the human form that it is impossible for it to express fear, anger, anxiety, anticipation, etc. We know in advance that no scientific discovery (e.g. that a fly has a brain) could justify the attribution of those forms of consciousness to flies. If a fly darts away from the sudden motion of something nearby, we are inclined to say that it was 'aware' of the motion. This is like saying that the motion 'frightened' it. This is all right. We might even speak of the fly's 'consciousness' of the motion. But we should realize that all this term refers to here is that tiny bit of behaviour. It is meaningless to wonder whether flies may have consciousness in a 'deeper' sense.

To say that the living human being is the paradigm for the application of psychological predicates and verbs, is to say that the *primary* application of these terms is to living human beings and that they are applied to other creatures by analogy. One might object that our application of some of these terms to dogs and cats is spontaneous: we do not apply them on the basis of explicit comparisons with people. True. So what is the justification for claiming that dogs can be said to be angry or frightened only by analogy with people?

The justification can be seen in the following way. Suppose there was a person, A, who has as good a mastery of psychological verbs and predicates in their application to people, in both first-person and third-person uses, as do any of us, but who always found it objectionable, absurd, comical—when other people applied these terms to lower animals. When the other people defend this application by pointing out similarities, he responds by pointing out genuine differences. The others (especially dog, cat, and horse lovers) would regard A with amazement and probably some indignation: 'Can't he see how *human* they are?' But no one could rightly say that A doesn't *understand* such terms as 'hungry', 'nervous', 'afraid', etc., since he gives ample proof that he does in his daily life with people. In contrast, if we try to imagine another person, B, who never learns to apply the ordinary mental expressions to *people*, but does astonish us by applying them to other animals just as we would, we could not say on the basis of this latter phenomenon that he understands these terms: his inability to apply them correctly to people would rule this out. B would have to be regarded as an idiot, whereas A could be thought of as intelligent and perceptive, but eccentric. The contrast between these two imaginary cases shows that the application of psychological terms to people is primary, and their application to other animals is by analogy.

Of a machine there is even less sense than in the case of a fly, for wondering whether it might be conscious or unconscious. This is easily seen from the fact that the terms 'dead' and 'alive' have no literal application to machines, but only a metaphorical application, as when we say 'The engine went dead' or 'The battery is dead' or 'In the nick of time the motor came alive'. Since 'alive' does not have a literal application to machines there cannot be a serious question as to whether a machine might literally be conscious or unconscious. There are machines that react to changes in colour of nearby lights, and it would not be surprising if some people began to speak of the machine as 'perceiving' the different colours. This would be a metaphorical way of saying that the colour changes activate some device in the machine. People invent or gradually fall into a new application of an expression. In this itself there is no harm; but it is easy to become confused by the appearance of the same word into thinking that the same thing is being said. Perhaps some day it will be said of some machines that they are sleeping or sleepy, in a daze or a trance, groggy, half-conscious, fully conscious, alert, witty, etc. What would these expressions mean when applied to machines? I have no idea. But their meaning when applied to machines will necessarily be different from their meaning when applied to people. One could not be saying the same thing in these two applications.

What can be said. Wittgenstein's remark that only of a living human being and what resembles one, can one say that it is conscious or unconscious, should not be construed as implying that the terms 'conscious', 'unconscious', are *always* applicable to living human beings. If my wife is and has been in good health, I should be taken aback if an acquaintance said to me, 'Is your wife conscious?' Certainly I would not reply, 'Oh yes, she is conscious'. This is not how these words are used: they cannot be said in such a case.

There is much confusion in philosophy about the notion of 'what can be said'. One may encounter the trivial retort that of course 'She is conscious' *can be said* in the above case—where this apparently means no more than that it is physically possible for a person to utter those words. Another retort, which probably comes to the same, is that though those words *are not said* in such a case, they *can* be said. In the present essay the expressions 'is (is not) said', 'can (cannot) be said', are being given exactly the same use. This is a familiar use of these expressions, as is illustrated by the use of similar expressions in games. A beginner in chess, who sought to avoid checkmate by removing his king from the board, might be told either 'That can't be done', or 'That isn't done', or 'That isn't a move'. The beginner's instructor might equally have said, 'Taking your king from the board makes no sense, has no meaning, in chess'. The beginner may have been familiar with another board game, somewhat similar to chess, in which a threatened piece can be removed from the board. The instructor might think that the beginner had confused chess with that other game. He might ask, 'What game do you think you are playing?' Similarly, in the situation where the question was put to me about my wife, if an observer knew that my wife was not and had not been ill, and knew that I knew this, he would be surprised by my saying, 'Oh yes, she is conscious'. He might afterwards ask, 'What were you saying, what did you mean?' He knows that 'She is conscious' is not said, cannot be said, in that situation; so he is perplexed as to what I was saying, what game I was playing. I take the question '*Can* those words *be said* in that situation?' as identical with the question 'Do those words *have an application* in that situation?' and also as identical with the question 'Do those words *make sense* in that situation?'

It is common for philosophers to have views or 'theories' about various concepts, theories which require that something can be said which in fact cannot be said. It might, for example, be a theory about the concept of *seeing*, a theory implying that a person cannot *see* a thing in front of him unless it *seems to him* that he sees it; or it might be a theory about *willing* and *acting*, a theory implying that whenever one *does* something one *tries* to do it; and so on. In many situations, however, where you see something (your hat or your car) it cannot be said that it *seems to you* that you see it; and in many cases where you *have done* something (gone for a walk or cashed a cheque) it cannot be said that you *tried* to do it. These theories are proved false by these implications. Why is this proved? Because the theories are supposed to be philosophical accounts or analyses of everyday concepts, concepts that are embedded in and reflected by the common use of language. A philosophical theory that contradicts this use is false, although it may have merits of another kind.

Philosophers usually want their theories to be true, and so considerable ingenuity has been spent on trying to establish that a philosophical account of a concept is not rendered false by being contrary to what can be said. A widely adopted defence is to distinguish between 'speech-conditions', which supposedly govern when it is appropriate to say or assert something, and 'truth-conditions', which are thought to govern when that which is said or asserted is true or false. Truth-conditions are sometimes regarded as identical with 'meaning-conditions', which are supposed to determine when a statement is meaningful. It is then argued that speech-conditions, and truth-or-meaning-conditions, are not the same; therefore it may be inappropriate *to say* something that is both meaningful and true. For example, it would have been inappropriate (because misleading) for me to have said 'My wife is conscious', yet if I had said it my statement would have made sense and would have been true.

The argument is specious. What is 'inappropriate to say' is being confused with what cannot be said. It might be inappropriate to make brilliant moves in a chess game with a beginner, because this would destroy his interest in learning the game; but the moves belong to chess, they can be made. The same holds for speech; it may on occasion be inappropriate (because rude, injudicious, unfair, misleading, etc.) to say something that can be said and is true (or false). But what cannot be said is neither true nor false; just as taking the king from the chess board is neither a correct move nor an incorrect move, but just not a move.

One source of the confusion is the failure to realize that 'what can be said' is a *relative* concept, relative to *particular situations*. Suppose the phone rings—you pick up the receiver, say 'Hello', and enter into conversation with the speaker at the other end of the line. Afterward it could be said that you answered the phone, not that you *tried* to answer it. If you couldn't get hold of the receiver, or dropped it breaking the phone, or there was no response from the other end, etc., then it could be said that you *tried* to answer the phone. There are indefinitely many different situations in which 'answered' can be said; similarly for 'tried to answer'; and indefinitely many situations in which it would not be clear that either thing could be said.

We are enormously attracted by the notion that it must be possible to formulate these nuances in principles that would give necessary and sufficient conditions for 'answered' and for 'tried to answer'—and generally for 'did' and 'tried to do'. We feel that if this were not so then the different examples of *trying* would not be united by 'a common character', and no one would be able to learn the use of the verb 'to try'. But the fact is that we are not aware of necessary and sufficient conditions, and it is question-begging to assume that therefore we must have an unconscious or inarticulate knowledge of such conditions. The same holds *a fortiori* for the idea that there must be necessary and sufficient conditions governing what can be said *in general*. There are absolutely no grounds for this assumption. Knowing a language is knowing what can be said and what cannot be said, *from case to case*. It is knowledge of particular cases, not of general rules.

A perplexing question is this: suppose that in a situation where a certain thing cannot be said, someone *says* it; what then? Consider again the example of your having answered the phone. Later in the day you are asked whether the phone rang and whether you answered it. You reply 'I *tried* to answer it'. What were you saying? I don't know. There is a tantalizing philosophical inclination to declare that when you said 'I tried to answer it', you were just saying that *you tried to answer it*!

Nothing could be more obvious! When you said '*p*', you were saying that *p*!

The emptiness of this 'explanation' is apparent. If one is puzzled as to what a person said or meant in uttering '*p*', it is no explanation to be told that in uttering '*p*' he meant *p*. A genuine explanation would mention some relevant circumstances of the situation. Suppose that in the past you had often been rude in answering the phone and had been rebuked for this: 'That is not how one answers the phone!' Then your reply, 'I *tried* to answer it', might have been an apologetic way of saying that you tried to answer the phone *politely*. But if you had never displayed a tendency to answer the phone rudely, and had never been reprimanded for this offence, then your reply did not mean *that*. Uncertainty as to what a person was saying when he uttered '*p*', may be removed by an expanded description of circumstances, both past and present—but not by the senseless insistence that in uttering '*p*' you were saying that *p*!

I am conscious. A more metaphysical objection to my claim that one cannot say that a person is conscious unless one believes that the person is or was or may be seriously ill, is the following: a conscious person is always aware of being conscious, and can always say or think 'I am conscious'; and if that person can say it, so can others say of him or her, 'He/she is conscious'.

But in what context would a person's saying 'I am conscious' be intelligible? If I have been injured in an accident, someone bending over me might say 'He seems to be unconscious', and I might murmur in response 'I am conscious'. My utterance would be intelligible in that situation. It would inform the other person that I am conscious *just by being a sign of consciousness*. Other signs would serve as well: opening my eyes, sitting up, *saying anything at all*. My utterance, 'I am conscious', would show the other person that I am conscious, just by being an utterance and not by virtue of the supposed 'propositional content' of the utterance.

In the absence of such a context couldn't I say 'I am conscious' to myself? Yes. But what would I be doing? Perhaps I had expected to be unconscious, or had *just emerged* from unconsciousness—so my saying to myself 'I am conscious!' might be an exclamation of relief or of joy: just as I might exclaim with joy 'I see!' on regaining my sight after a period of blindness; or as I might say in triumph 'I am here' after overcoming severe obstacles to reach my destination. These utterances have that sort of meaning in those contexts. We should not suppose that when 'I am conscious!' is an exclamation of surprise or joy it also expresses a *perception* of one's condition.

There is an inclination to think that this cannot be right: 'If I exclaim 'I am conscious!', expressing my amazement and relief, then surely I must *perceive* (be aware) that I am conscious—for how could I be amazed that I am conscious unless I perceived it?' This appears to be a powerful consideration—until we reflect on the idea of one's perceiving that one is conscious—as Wittgenstein does when he asks, 'Do I observe myself, then, and perceive that I am seeing or conscious?' And again: 'But isn't it a particular experience that occasions my saying "I am conscious again"?—*What* experience?'[36] When we reflect on these questions we realize that if my being conscious could be something I experienced, if I could perceive that I am conscious—then couldn't I also perceive that I am *un*conscious? And also, if it is possible for me to perceive that I am conscious, shouldn't an error of perception be possible, namely, my seeming to myself to perceive that I am conscious, when actually I am unconscious? The absurdity of these implications clearly shows the nonsensical character of the notion that I might perceive that I am conscious.

Strangely enough, there is an inclination to think that 'I am conscious' or 'I have consciousness' is the most fundamental proposition there is—even more fundamental than 'I exist'. Of course one has this inclination only when doing philosophy, e.g. searching for 'the foundations of human knowledge'. But instead of its being a fundamental proposition, is it even a 'proposition'? Wittgenstein imagines its being said: 'Nothing is so certain as that I possess consciousness'—and also its being said: '"I have consciousness"—that is a statement [*Aussage*] about which no doubt is possible'. About this latter remark Wittgenstein asks: 'Why should not that say the same as: "'I have consciousness' is not a proposition" [*ist kein Satz*]?'[37]

[36]*PI*, §417.
[37]*Zettel*, §401, §402.

'I have consciousness' is certainly a *sentence*. What could be meant by questioning whether it is a 'proposition' or a 'statement'? Let us note the following points: first, the sentence 'I have consciousness' cannot be used to report a perception or observation, as we have already seen; second, the sentence does not have a meaningful negation; third, although in some circumstances my uttering the sentence could assure someone that I am conscious, this would be due solely to my *uttering* the sentence, not to any 'propositional content' of the sentence. What these points show is that the sentence 'I am conscious' does not have a fact-stating use: it does not report or describe the speaker's condition. This might be what Wittgenstein meant by suggesting that it is not a 'proposition'.

The sentences 'I am conscious', 'I have consciousness' and their negations do have a genuine use as antecedents of conditional statements: e.g. 'When I am conscious after the operation, please ask my brother to come to the hospital'. But as first-person, present-tense, simple-indicative sentences, they could be eliminated from language without any loss. This is in striking contrast with the third-person sentence 'He/she is conscious' which can be used to express an observation, a hypothesis, or a conjecture. The simple-indicative 'I am conscious' is a sentence without a job. It provides a good illustration for Wittgenstein's remark:

> It is not every proposition-like formation (*satzartige Bildung*) that we know how to do something with . . .; and when we are tempted in philosophy to count some quite useless thing as a proposition, that is often because we have not sufficiently considered its application.[38]

Aware of being conscious. The thought that one does not perceive one's own consciousness, may produce acute discomfort. 'If I don't perceive my own consciousness, how am I able to be aware of being conscious?' But is there such a thing as being aware of being conscious? When would one say such a thing? It is possible to think of a situation in which it might be said. Suppose that a friend had recently undergone surgery under general anaesthetic. You might ask her, 'Did you recover

[38] *PI*, §520.

consciousness gradually or suddenly?' She might reply: 'It was a remarkable experience. I was suddenly aware of being fully conscious!' But she might have said instead: 'It was a remarkable experience. I was suddenly fully conscious!' Considering her remark as information the words 'aware of being' add nothing. (They do contribute dramatic emphasis.) The speaker was not contrasting 'aware of' with 'not aware of'—for it would have been nonsense for her to have said, 'I was suddenly fully conscious but was not aware of it'. This speaker might have meant that not only did she suddenly become fully conscious, but also said to herself, 'I am fully conscious!', perhaps with a feeling of astonishment. If this is how we are to understand her testimony that she was 'aware' of being conscious, then it is clear that since people are not constantly, or hardly ever, saying or thinking this to themselves, there is no support here for the idea that whenever one is conscious one is aware of it.

Consciousness perceived from without. We have been reflecting on the concept of consciousness, both consciousness *of* something, and consciousness *tout court*. A prominent philosophical tradition holds that states, events, episodes of consciousness, are only perceived by a person from within and never by an observer of the person. The truth is just the reverse of that. To put it somewhat tendentiously, the phenomena of consciousness are perceived from without rather than from within.

Among philosophers currently reflecting on the science of psychology there is much concern over the question of whether psychology is capable of dealing with what is variously called, 'consciousness', 'the subjective character of experience', 'the qualitative character of mental states', 'qualia'. Thomas Nagel says:

> We have at present no conception of what an explanation of the physical nature of a mental phenomenon would be. Without consciousness the mind–body problem would be much less interesting. With consciousness it seems hopeless.[39]

Ned Block says:

[39]Thomas Nagel, 'What Is It Like to Be a Bat?', in *Readings in Philosophy of Psychology*, edited by Ned Block, two volumes, Methuen, 1980, vol. 1, p. 159. (This volume will heareafter be referred to as Block vol. I.)

> I do not see how psychology in anything like its present incarnation could explain qualia . . . it looks as if qualia are not in the domain of psychology.[40]

Wittgenstein remarks that it is a 'misleading parallel' to say that 'psychology treats of processes in the psychical sphere, as does physics in the physical'; and he goes on to say:

> Seeing, hearing, thinking, feeling, willing, are not the subject of psychology *in the same sense* as that in which the movements of bodies, the phenomena of electricity, etc., are the subject of physics. You can see that from the fact that the physicist sees, hears, thinks about, and informs us of these phenomena, and the psychologist observes the *external reactions* (the behaviour) [*die Äusserungen* (*das Benehmen*)] of the subject.[41]

Does not Wittgenstein's remark agree with the view of Nagel and Block that empirical psychology can only deal with the physiological processes and physical behaviour of subjects, and not with the phenomena of mind and consciousness? No. This can be seen from a further remark:

> Then psychology treats of behaviour, not of the mind? What do psychologists record—What do they observe? Isn't it the behaviour of human beings, in particular their utterances? But *these* are not about behaviour.[42]

A human subject of a psychological experiment will *say* various things: e.g. 'The light seems brighter now', 'I am beginning to feel slightly faint', etc. These utterances are reports, not of the subject's behaviour, but of his sensations. So the subject's *sensations*, not just his verbal behaviour, become part of the data of the experiment. These reports of the subject would be about changes in *consciousness*, changes in 'the qualitative character of experience'. Thus, consciousness and 'qualia' are clearly within the domain of psychology. An experiment might demonstrate causal connections between physical changes in the environment

[40] Ned Block, 'Troubles with Functionalism', in Block vol. I, p. 289.
[41] Wittgenstein, *PI*, §571.
[42] *PI*, p. 179.

or in the subject's body, and changes in the subject's sensations. Contrary to Nagel, this would be an intelligible 'explanation' of the physical nature of a mental phenomenon'.

And it is not just *utterances* that are expressive of sensation, feeling, emotion, disbelief. A narrowing of the eyes, a tightening of the mouth, a shrug, the paling of the countenance, can display suspicion, or a stiffening of the will, or fear. As Proust says, 'the truth has no need to be uttered to be made apparent, and one may perhaps gather it with more certainty, without waiting for words and without even taking any account of them, from countless outward signs.' Thus it is not easy to understand why Nagel and Block think that movements of consciousness or qualities of feeling are not discernible or explicable through the observation of those 'outward signs'. In section 2 we shall examine their position.

Behaviour in circumstances. Let us review what we have said about consciousness, both transitive and intransitive. We recall Brentano's two assertions, first, that 'no mental phenomenon is perceived by more than a single individual', second, that it is a 'general characteristic of all mental phenomena that they are perceived only in inner consciousness'. Brentano seems to be making the following two claims: first, that no person other than myself ever can perceive that I am frightened, angry, intend to do something, or am conscious of anything whatever; second, that whenever I am frightened, angry, etc., I myself do perceive this. These two claims have got things exactly backwards. Others often perceive my feelings, intentions, thoughts. How are they able to do this? By noting my facial expressions, gestures, movements, and hearing my exclamations, remarks, questions. Roughly speaking, they know about me from my *behaviour*.

This rough way of speaking can be misleading. For it is not my behaviour observed narrowly, that can yield this knowledge of my feelings and thoughts, but this behaviour *in certain circumstances*. Suppose that at a gathering I turn away *in disgust* from a group of people with whom I have been conversing. A friend of mine observes this. How could he? Might not that departure have been from boredom, not disgust? How could my friend know *which* it was? Well, he heard what the others in the group were saying, and he knew enough about my convictions or prejudices to realize that those remarks would be for me *an occasion for disgust*. Another observer might have seen my abrupt movement away from the group, but have failed to perceive it *as* an expression of disgust—because he didn't hear those remarks, or heard them but didn't understand them, or didn't know that I had those strong convictions.

Brentano's second claim is that if I was disgusted I must have perceived my disgust. This is partly false and partly not intelligible. It is false, in that I might not have understood my action. If questioned about it immediately afterwards I might have been unable to give a true explanation of my rude departure, perhaps because reflection had never made me sufficiently aware of those sentiments of mine that were offended: someone who knew me well and who perceived my action, might have had a better understanding of it than I did. But in another way Brentano's second claim is not so much false as obscure. Suppose I knew perfectly well that I had turned away in disgust: I 'knew' this *in the sense* that immediately after my action I could readily give a correct explanation of it. What would it mean to say, with Brentano, that I *perceived* my disgust at the moment I turned away in disgust? Well, it might mean that at that moment I thought to myself, 'These remarks disgust me!' Indeed that might have happened: but also it might *not* have happened. In the latter case, what would be that phenomenon which, according to Brentano, was not just my disgust but my *perception* of my disgust? I don't know. This is the unintelligible aspect of Brentano's claim. Only in some situations can it be said that a person perceives (is aware of) his/her own feeling of irritation, contempt, disgust, and so on (including all forms of transitive consciousness).

With regard to intransitive consciousness or unconsciousness, there is a more obvious nonsense in speaking of one's being aware of being conscious or unconscious. One source of the feeling that a conscious person *must* be aware of being conscious may be the realization that 'He was not aware of being conscious' would be a strange thing to say. Strange indeed! It is easy to think of a case where we could say that A is not aware of B's being conscious: a patient in surgery lost consciousness but a few minutes later regained consciousness; the surgeon, whose back was turned, did not see the indications of regained consciousness, and so was not aware of the patient's being conscious. But in what circumstances should we say that the patient was not aware of being conscious? I don't know. Should we say it if the patient muttered 'I am unconscious' or 'I am not aware of being conscious' or said, after recovery from the operation, 'I was conscious but was not aware of it'? Certainly not. It appears that 'He is/was not aware of being conscious' is a sentence that we do not want to apply to *any* situation. But this is no support for the idea that a conscious person must be aware of being conscious. Quite the opposite! For being aware of being conscious is supposed to contrast with not being aware of being conscious: if this latter is a nothing the contrast is a nothing too.

Self-consciousness. I have not mentioned self-consciousness. What is it? There is a common use of 'self-conscious' according to which being self-conscious is being embarrassed, shy, awkward, when noticed by others. This is not relevant to our inquiry. There is also a philosophical notion of 'self-consciousness' that is supposed to indicate each person's consciousness of a unique object, different for each of us, which is what the pronoun 'I' stands for. I agree with Anscombe that the word 'I' is not used to designate any object, and that this philosophical notion is illusory.[43]

Anscombe goes on to declare that 'self-consciousness' is 'something real'. She says: 'The expression "self-consciousness" can be respectably explained as "consciousness that such-and-such holds of oneself".'[44] What she has in mind under the heading of 'self-consciousness' is the normal ability of any person to report his/her bodily postures, movements, sensations, intentions—such reports being generally true, although *not* based on observation. The facts to which Ans

[43] G. E. M. Anscombe, 'The First Person', in *The Collected Papers of G. E. M. Anscombe*, Blackwell, 1981, vol. II. Also: Norman Malcolm, 'Whether "I" Is A Referring Expression', *Intention & Intentionality*, edited by Cora Diamond and Jenny Teichman, Harvester Press', 1979.
[44] Anscombe, 'First Person', p. 26.

combe thus calls attention are important. One can invoke the expression 'self-consciousness' as a technical term in philosophy to refer to these facts. But it should be noticed that the term is not used this way in ordinary language: not *every* normal human being is said to be 'self-conscious': distinctions are drawn between people in terms of whether they are, or are not, 'self-conscious'.

In everyday conversation, and in literature, a person is said to be 'self-conscious' who *reflects* a lot on his own attitudes, interests, personality, and on his reactions to other people and to situations in human life, and also on his responses to music, literature, art and nature. People who are self-conscious in this sense, vary greatly in the *degree* of their self-consciousness. Such self-consciousness is often stimulated and nourished by the observations of others about one's self. One's own person becomes an object of one's study.

2 The Subjective Character of Experience

Thomas Nagel says that currently popular programmes in the philosophy of mind, such as materialism, physicalism, functional analysis, fail to give an account of *consciousness*. He identifies consciousness with 'the subjective character of experience'. He says this is something that is

> not captured by any of the familiar, recently devised reductive analyses of the mental, for all of them are logically compatible with its absence. It is not analyzable in terms of any explanatory system of functional states, since these could be ascribed to robots or automata that behaved like people though they experienced nothing. It is not analyzable in terms of the causal role of experiences in relation to typical human behaviour—for similar reasons. I do not deny that conscious mental states and events cause behaviour, nor that they may be given functional characterizations. I deny only that this kind of thing exhausts their analysis.[1]
>
> If physicalism is to be defended, the phenomenological features must themselves be given a physical account. But when we

[1] Nagel, 'What Is It Like to Be a Bat?', p. 160.

> examine their subjective character it seems that such a result is impossible.[2]

This something that physicalist or functionalist theories allegedly cannot deal with is not only called 'consciousness', or the 'subjective' or 'phenomenological' character of experiences: Nagel also alludes to it in the phrase 'what it is like'. Nagel assumes that bats, for example, have experiences, and he says that 'the essence of the belief that bats have experience is that there is something that it is like to be a bat'.[3] Nagel believes that a human being cannot understand what it is like to be a bat.

> We believe that bats feel some versions of pain, fear, hunger, and lust, and that they have other, more familiar types of perception, besides sonar. But we believe that these experiences also have in each case a specific subjective character, which it is beyond our ability to conceive.[4]

> In contemplating bats we are in much the same position that intelligent bats or Martians would occupy if they tried to form a conception of what it was like to be us . . . we know they would be wrong to conclude that there is not anything precise that it is like to be us . . . because we know what it is like to be us.[5]

'What it is like' is an interesting phrase. Suppose I sat next to a long distance lorry driver at a lunch counter and we struck up a conversation. I might say to him, 'What is it like to be a lorry driver?' This would be a vague question, but at least it would be an invitation for him to say *something* about being a lorry driver. He might reply, 'What do you mean?', thereby asking me to be more specific. I might have something more specific in mind—whether it is difficult to manœuvre a huge lorry; whether he finds his job boring, challenging, satisfying; whether he resents having to spend so many nights away from home. On the other hand I might have nothing so definite in mind: I might just want him to relate some of his experiences on the road.

[2]Ibid., p. 160. [3]Ibid., p. 161. [4]Ibid. [5]Ibid., p. 162.

The mention of *experiences* brings us nearer to Nagel's topic, but still not near enough: for Nagel speaks not just of 'experiences' but of their 'subjective character'. This suggests that if the lorry driver told me of an experience of his on the road—for example, that he was marooned in a blizzard for twenty-four hours—he might not have, so far, disclosed anything about its 'subjective character', of 'what it was like'. Suppose I ask, 'What was it like?', and he replies: 'I was terribly cold, hungry, thirsty, and afraid that I would freeze to death before help came'. But if a bat's pain, fear, or hunger has a 'special subjective character', presumably this would also be true of the driver's feeling cold, hungry, and fearing he would freeze to death. According to Nagel I am unable to conceive of the 'subjective character' of a bat's fear or hunger: but with the driver there is no problem of inconceivability, since he is a human being and 'we know what it is like to be us'. So I understand the 'subjective character' of the driver's feeling cold.

But is this so? Is it something that can be described? Can I even in my own case put into words the 'subjective character' of feeling cold? Recalling an occasion when I was exceedingly cold I might supply further details, such as that my feet felt 'like blocks of ice', or that my fingers were numb with cold. On Nagel's view, however, won't any detail of experience have its own 'subjective character'? Won't there be a further question of 'what it was like for me' for my fingers to feel numb with cold? And what can I say about that? Is the 'subjective character' of that experience something that cannot be expressed?

Nagel thinks that the problem of inconceivability arises not only with bats and other creatures, but also with some human beings who are abnormal in certain ways:

> The subjective character of the experience of a person deaf and blind from birth is not accessible to me . . . nor presumably is mine to him. This does not prevent us each from believing that the other's experience has such a subjective character.[6]

Since I have sight and hearing I am supposed to know what seeing and hearing are like—to know their subjective character. But what is it that I am supposed to know? I close my eyes and

[6]Nigel, 'What Is It Like to Be a Bat?', p. 162.

see nothing. I open them and see my desk, typewriter, books and bookshelves. But I have merely enumerated some of the things I see. I have not described *seeing itself*. Is it that I *know* what it is like but cannot describe it?

I can tell from another person's behaviour whether he is sighted or blind. And from the behaviour and circumstances of a person whom I know to be sighted, I can often tell whether at the moment he sees or doesn't see: (perhaps the room is dark or he is blindfolded). Although I often do know that another person sees the objects in front of him, do his behaviour (including what he says) and the physical circumstances, inform me of the 'subjective character' of his seeing? Apparently not. For Nagel says that consciousness (or the subjective character of experience) 'is what makes the mind – body problem really intractable'.[7] Now if I could determine the subjective character of another's seeing, from observing his behaviour and circumstances, there would be nothing 'intractable' about it.

If I cannot produce a description of the subjective character of seeing can't I at least *exhibit* that character *to myself*? How would I do that? Would I look at something and say to myself, 'Seeing is *this*'? But this *what*? If I were to exhibit to myself the subjective character of hearing, or of pain, or of hunger, or of fear, or of feeling cold—wouldn't the exhibition each time include the thought 'It is *this*'? But surely the *this* of seeing must be different from the *this* of hearing, or pain, or hunger. Yet I do not know how to display to myself the differences in 'subjective character' of these different experiences. Could it be that the 'subjective character' of seeing, hearing, pain, and of all other experiences, is the *same*?

One might object that seeing must be different from hearing. Of course it is. The organs of sight are different from the organs of hearing; the behavioural expressions of seeing and hearing are different; the language of seeing is different from the language of hearing. But our concern is with the *subjective character* of seeing and hearing. Can I be sure that *that* is the same, or is different, in the two cases? Might not *that* even be sometimes the same and sometimes different?

[7]Nigel, 'What Is It Like to Be a Bat?', p. 159.

These questions show that I don't know what I am talking about. When I try to fix my attention on my seeing or hearing or pain and think 'It is *this*', not only don't I know whether *others* do or don't have *this*—I don't even know, in my own case, whether my present *this* is or isn't *the same* as a previous *this*. Not only cannot I display the 'subjective character' of seeing or hearing to others; I cannot even display it to myself.

In *Remarks on the Philosophy of Psychology* Wittgenstein comments on the notion of the 'content' of experience: but it is clear that he is referring to what Nagel calls the 'subjective character' of experience—'what it is like'.

> The 'content' of experience, of experiencing: I know what toothaches are like, I am acquainted with them, I know what it's like to see red, green, blue, yellow, I know what it's like to feel sorrow, hope, fear, joy, affection, to wish to do something, to remember having done something, to intend doing something, to see a drawing alternately as the head of a rabbit and of a duck, to take a word in one meaning and not in another, etc. I know how it is to see the vowel *a* grey and the vowel *u* dark purple.—I know, too, what it means to parade these experiences before one's mind. When I do that, I don't parade kinds of behaviour or situations before my mind.—So I know, do I, what it means to parade these experiences before one's mind? And what *does* it mean? How can I explain it to anyone else, or to myself?[8]

Here Wittgenstein first gives vent to his inclination to say that he 'knows what it's like' to see red, feel sorrow, etc.; but then he rounds on this inclination. He is supposed to be able to bring before his mind the 'content' of those experiences—not the behaviour or the circumstances, but the 'inner character' of seeing red or feeling sorrow. Yet he realizes that he doesn't know what the 'content' or 'inner character' is that he is supposed to pass in review before his mind.

In another passage Wittgenstein describes the inclination to try to give oneself an exhibition of the 'content' of experience:

> The *content* of experience. One would like to say 'I see red *thus*',

[8]Ludwig Wittgenstein, *Remarks on the Philosophy of Psychology*, vol. I, edited by G. E. M. Anscombe and G. H. von Wright, translated by G. E. M. Anscombe, Blackwell, 1980, §91.

> 'I hear the note that you strike *thus*', 'I feel sorrow *thus*', or even '*This* is what one feels when one is sad, *this*, when one is glad', etc. One would like to people a world, analogous to the physical one, with these *thus*es and *this*es. But this makes sense only where there is a picture of *what is experienced*, to which one can point as one makes these statements.[9]

These *thus*es and *this*es are impotent. They could be genuine exhibitions only if they were accompanied by 'pictures', i.e. *representations*, of what is experienced. Representations that one could show to another person as well as to oneself. But this attempt to concentrate on the 'content' or 'subjective character' of experience, an attempt that excludes any employment of representations of what is experienced, does not exhibit or explain anything. It is a purely metaphysical use of these terms.

The origin of this metaphysical notion of 'content' or 'subjective character' or 'what it is like', is described by Wittgenstein as follows:

> Where do we get the concept of the 'content' of an experience from? Well, the content of an experience is the private object, the sense-datum, the 'object' that I grasp immediately with the mental eye, ear, etc. The inner picture.[10]

One wants to give oneself an ostensive definition that is 'private' or 'inner'. The feeling is that there is *something* on which one can focus one's attention, but *cannot* exhibit to anyone else. But the word 'cannot' is misused here. The 'It is *this*', accompanied by an 'inward glance', but not accompanied by any sample, illustration, or description, does not pick out anything at all. It is not the case that it picks out something that 'cannot' be exhibited to others. As Wittgenstein says in the *Investigations*:

> The great difficulty here is not to represent the matter as if there were something one *couldn't* do. As if there really were an object, from which I derive its description, but I were unable to show it to anyone.[11]

[9]Ibid., §896. [10]Ibid., §109.
[11]Wittgenstein, *PI*, §374.

The metaphysical notion of consciousness. Recall Nagel's assertion that consciousness makes the mind–body problem intractable. How is he using the word 'consciousness'? Surely not in the sense in which consciousness is contrasted with unconsciousness, as when we say that a person has 'regained consciousness'. When someone has been knocked unconscious we know what are the signs of returning consciousness. There is nothing intractable about consciousness in *this* sense. It would seem that the notion of 'consciousness' that Nagel finds intractable is the metaphysical notion which Wittgenstein characterizes in the *Investigations* when he speaks of

> The feeling of an unbridgeable gulf between consciousness and brain-process . . . When does this feeling occur . . . It is when I, for example, turn my attention in a particular way on to my own consciousness, and astonished, say to myself: THIS is supposed to be produced by a process in the brain!—as it were clutching my forehead.—But what can it mean to speak of 'turning my attention to my own consciousness'? This is surely the queerest thing there could be! It was a particular act of gazing that I called doing this. I stared fixedly in front of me—but not at any particular point or object. . . . My glance was vacant . . .[12]

Notice that Wittgenstein at first speaks as if there really was such a thing as 'turning my attention on to my own consciousness'—but when he reflects on what goes on in such a case, what he finds is an act of vacant gazing, which occurs along with the thought of a gulf between consciousness and brain-process. That mysterious consciousness is merely a vacant gazing, where one's attention is fixed on *nothing*.

If this is the notion of 'consciousness' that puzzles Nagel then he has understated his problem. He thinks his problem is *not* that of the so-called 'privacy' of experience:

> I am not adverting here to the alleged privacy of experience to its possessor. The point of view in question is not one accessible only to a single individual. Rather it is a *type*. It is often possible to take up a point of view other than one's own, so the comprehension of such facts is not limited to one's own case.

[12] Ibid., §412.

> There is a sense in which phenomenological facts are perfectly objective: one person can know or say of another what the quality of the other's experience is.[13]

But *can* one person know what the 'quality' of another's experience is? Not if this means knowing whether the other has THIS—where no further specification is provided, and where one's attention is not directed on anything. Of course there is a familiar sense of describing the 'quality' or 'subjective character' of one's experience. Was not the lorry driver describing something of the 'subjective character' of his experience in the blizzard, when he said that he feared he would freeze to death? But since, on Nagel's view, *every* experience has a 'subjective character', the subjective character is not reached by any description of experience: it lies beyond every description. No wonder the 'inner quality' or 'subjective character' of experience is baffling. This is because Nagel's use of these expressions is *regressive*: it has a constant backward movement. This helps to give him the impression that he is referring to something when he is referring to nothing. Given *any* experience, Nagel is saying that *that* experience too has a 'specific subjective character'. The regressive feature of his use of the expression the 'subjective character' is what gives rise to the feeling that he is talking about something that is hard to get at. This is an example of what Wittgenstein diagnoses as interpreting 'a grammatical movement made by yourself as a quasi-physical phenomenon which you are observing'.[14] In Nagel's regressive, metaphysical employment of the phrase, the 'subjective character' of an experience of yours is absolutely inaccessible to me. This is because no description or illustration of its 'subjective character' is forthcoming, and therefore no *comparison* of yours and mine is possible.

Knowing what it is like. Do I know what it is like for a person to be red-green colour blind from birth? I know that he doesn't see any difference between reds and greens: *that is*, he doesn't discriminate between them, and he says they look the same. Suppose I wonder whether it is that greens look red to him or

[13] Nagel, 'What Is It Like to Be a Bat?', p. 163.
[14] Wittgenstein, *PI*, §401.

that reds look green? He cannot tell me: for one aspect of his red-green colour blindness is that he is not able to apply the expressions 'looks red' and 'looks green' in the way that people with normal colour vision do. It is useless to ask *him* whether reds look green or greens look red. There is an inclination to say: 'Still, isn't it *possible* that reds and greens all look red to him?' What is this 'possibility' supposed to mean? Is it that if I received his sense-impressions then both reds and greens might look red to me? But how could this disclose what *his* colour impressions were? 'Knowing what colour blindness is like' has an intelligible sense in which it just means, knowing that someone cannot discriminate between certain colours.

Do sighted people know what it is like to see? We have an inclination to say that they *do* and that the congenitally blind *do not*. But both can know a lot about the behavioural differences between the blind and the sighted—for example, that it will make a great difference to the way the sighted walk if they are blindfolded, whereas it will make little or no difference to the blind.

'But the blind don't know what the "inner experience" of seeing is.' Well, do the sighted know what it is? Do I know? If I were asked to describe the 'inner experience' of seeing I wouldn't know what to say. If I said 'It is THIS', that would not explain anything. What I can describe is how the lives of the blind are different in many ways from those of the sighted.

Nagel broaches the curious suggestion that there might eventually be developed 'an objective phenomenology' the goal of which 'would be to describe, at least in part, the subjective character of experiences in a form comprehensible to beings incapable of having those experiences'. 'One might try, for example, to develop concepts that could be used to explain to a person blind from birth what it was like to see'.[15] But the idea of teaching someone what seeing is, is completely opaque—unless it means teaching him (if he doesn't already know it) the use of the word 'see'. *Some* of the use of this word can be taught to a blind person: but much of it cannot—that is implied by his being blind. If he is blind he cannot move about in his environment as sighted people do—but also he does not have

[15] Nagel, 'What Is It Like to Be a Bat?', p. 166.

the full use of the language of vision that sighted people have. One is inclined to think that to know what seeing is would require a knowledge of the 'inner quality' of seeing. But this 'requirement' is not intelligible.

What would show that a blind person *understood* an explanation of 'what it is like to see'? One can explain the moves of chess pieces to someone who doesn't know chess. That he understands the explanation will be shown by his making correct moves on his own. So can the blind person show his understanding of the description of what it is like to see, by making true statements about what he sees and of how things visually look to him? 'But that is absurd—for then he would no longer be blind!' Of course it is absurd. But the absurdity exposes an important point, namely, that nothing intelligible is meant by 'understanding what it is like to see' *other than* just having a normal mastery of the language of vision. It isn't as if this latter were one thing and that understanding what seeing is like were an additional thing.

The bafflement that one feels about the undertaking of trying to describe the 'inner quality' or 'subjective character' of seeing, arises from the impression that a sighted adult not only displays normal visual discrimination, and a normal use of the word 'see', but *also* knows what seeing is like. As if there were tucked away inside seeing, the inner quality of what it is like. But in so far as 'knowing what seeing is like' has any meaning at all, it refers to nothing other than the ability that a sighted adult has of making visual discriminations, reports and judgements. This accounts for Nagel's confidence that a sighted adult 'knows what it is like to see': for what this amounts to is the tautology that such a person is not only able to make visual discriminations, but also can employ the language of sight in the normal way. It would make some sense (although it would sound queer) to say of a sighted child who hasn't yet learned any language, that it doesn't 'know what seeing is'. But it would make no sense to speak of *teaching* it what seeing is, unless this just meant—teaching it to employ the word 'see' in the ordinary way. This could not be done with a blind child. Thus it is meaningless to suppose, as Nagel does, that we might 'develop concepts that could be used to explain to a person blind from birth what it was like to see'.

Functionalism. The protest that Nagel made against some programmes in the philosophy of mind is presented in sharper focus in a current controversy about 'functionalism'. What is 'functionalism'? It is the theory that a 'mental state' is a 'functional state'. What is a functional state? It is a state of an organism, or of a machine, which is definable in terms of *causal* relations, i.e. relations to 'inputs, outputs, and successor states'. The view is that the concept of a mental state is the concept of a state that has a certain 'causal role'. As David Lewis puts it,

> Our view is that the concept of pain, or indeed of any other experience or mental state, is the concept of a state that occupies a certain causal role, a state with certain typical causes and effects.[16]

On this view, presumably, the difference between the 'mental states' of remembering and expecting would be that the one has certain typical causes and effects and the other has different typical causes and effects. In section 3 I will be examining this causal theory; but now I wish to consider a criticism directed against it by some philosophers, i.e. the criticism that functionalist definitions of mental states cannot account for the 'qualitative character' ('phenomenological character', 'raw feel', 'qualia') that at least some mental states are thought to possess.

This criticism closely resembles Nagel's charge that functionalist analyses cannot deal with the 'subjective character' of experience. Nagel seems to use the terms 'experiences' and 'conscious mental states' interchangeably. It is likely that those 'mental states' that are said to have 'qualitative character' are identical with those 'conscious mental states' that are said by Nagel to possess 'subjective character'. Nagel did not make clear what the 'subjective character' of experience was supposed to be. It will not be surprising if it turns out the same with regard to the 'qualitative character' of mental states.

[16]David Lewis, 'Mad Pain and Martian Pain', in Block vol. I, p. 218.

Ned Block and Jerry Fodor are among those who contend that functionalism cannot deal with the 'qualitative character' that some mental states allegedly have. They say:

> It does not . . . seem entirely unreasonable to suggest that nothing would be a token of the type 'pain state' unless it felt like a pain, and that this would be true even if it were connected to all the other psychological states of the organism in whatever way pains are.[17]

Do these remarks give a clue as to what is meant by the 'qualitative character' of pain? Nothing, it is said, would be a pain unless 'it felt like a pain'. Does every pain 'feel like a pain'? Does anything other than a pain 'feel like a pain'? Is the suggestion that nothing would be a 'pain state' unless it felt like a pain, just a cumbersome way of saying that a person is in pain if and only if he feels pain? This is true, but it doesn't tell us in what way the 'qualitative character' of pain differs from just pain.

Absent qualia. According to a functionalist theory called 'the functional state identity theory' (FSIT) a mental state, P, and a mental state, Q, are of 'the same type' if P and Q are 'functionally identical'—that is, if P and Q are 'identically connected with inputs, outputs, and successor states'.[18] Block and Fodor say that a serious difficulty for FSIT is the possibility of 'absent qualia'. They state the 'absent qualia' problem as:

> For all that we now know, it may be nomologically possible for two psychological states to be functionally identical (that is, to be identically connected with inputs, outputs, and successor states), even if only one of the states has qualitative content. In this case, FSIT would require us to say that an organism might be in pain even though it is feeling *nothing at all*, and this consequence seems totally unacceptable.[19]

Certainly this would be an absurd consequence, if it is a consequence of FSIT. But we have not yet seen what the distinction is between pain and the 'qualitative content' of pain.

[17]Ned Block and Jerry Fodor, 'What Psychological States Are Not', in Block vol. I, p. 244. [18]Ibid., p. 245. [19]Ibid.

Sydney Shoemaker offers a limited defence, not of FSIT specifically, but of functionalism ('broadly construed') against the 'absent qualia' objection. He agrees with Block and Fodor that some mental states have 'qualitative character', and he chooses to call a state a 'qualitative state' just in case it has 'qualitative character'.[20] He offers the following example of a qualitative state:

> There is a qualitative state someone has just in case he has a sensation that feels the way my most recent headache felt.[21]

Presumably, 'feeling the way his most recent headache felt' was the qualitative character of Shoemaker's most recent headache. Notice that Shoemaker does not tell us *how* his headache felt. He has not told us what *was* the 'qualitative character' of that headache.

There is something curiously evasive in the manner in which the philosophers we are discussing speak of the 'qualitative character' of a mental state. We noted the same shyness in Nagel's reference to the 'subjective character' of seeing or of fear; he didn't say what that was. Similarly, when Block and Fodor say that nothing would be a 'pain state' unless 'it felt like a pain', they do not go on to explain what 'feeling like a pain' is, or how it differs from just being a pain.

Do all pains feel the same? Of course not. There are many different characterizations of the 'quality' of pain: throbbing, piercing, burning, gnawing, dull, etc. These are *genuine* qualitative characterizations of pain. If we supplied one of these characterizations (e.g. 'piercing'), instead of the obscure phrase 'feeling like a pain', it would be obviously false that nothing is a pain unless it feels piercing. In the absence of any genuine qualitative description, to say that nothing is a pain unless it 'feels like a pain' seems to say no more than that nothing but a pain is a pain.

And why does Shoemaker too speak so vaguely of 'a sensation that feels the way his headache felt'? What *was* the way it felt? There is not just one way that headaches feel. Some are pulsating, some are dull, some grip one's skull like an iron band. Shoemaker fails to give us a qualitative description of his

[20]Sydney Shoemaker, 'Functionalism and Qualia', in Block vol. I, p. 253.
[21]Ibid.

headache. Therefore his example does not help us to understand what the 'qualitative character' of a mental state is supposed to be.

Shoemaker goes on to say that 'being in pain typically gives rise, given appropriate circumstances, to what I shall call a "qualitative belief", i.e. a belief to the effect that one feels a certain way . . .'[22] One feels a certain way! *What* way? Depressed, irritable, sorry for oneself? Being in pain sometimes results in these feelings, but not 'typically'.

It would appear that the 'qualitative belief' to which being in pain 'typically gives rise' is the 'belief' that one is in pain. Of course it is a misuse of language to say that a person 'believes' (or 'doesn't believe') he is in pain. But that isn't the present point—which is that Shoemaker is not disclosing what the 'qualitative character' of being in pain *is*.

Shoemaker concedes that it is difficult 'to distinguish pain from its qualitative character'.[23] He believes that it will be easier for him to clarify this supposed distinction by referring to 'visual experience'.[24] He begins by saying that 'If I see something, it looks somehow to me . . .' This is a curious remark. It is often true that when I see a man in the street there is a 'somehow' that he looks to me, e.g. tired, or bored, or shabby, or suspicious, etc. But also it is often true that I would not know what was meant, if someone asked 'How does that man look to you?' Furthermore, if I am walking in crowded streets there may not be *any* particular *look* that all or some of those people present to me. Certainly they do *not* 'look like people': only in special circumstances (e.g. dim outlines in heavy fog, or tiny figures at a great distance) would people 'look like people' to me. The idea that anything I see must 'look somehow' to me, suggests that we are in sense-datum country.

We are still hoping to understand what is meant by the 'qualitative character' of a mental state. An example would be helpful. Shoemaker now offers one. He says:

> Being appeared to in a certain way, e.g. things looking to one the way things now look to me as I stare out of my window, I

[22]Ibid., p. 254. [23]Ibid., p. 256. [24]Ibid.

> take to be a qualitative state. So seeing essentially involves the occurrence of qualitative states.[25]

Remember that Shoemaker has defined 'a qualitative state' as a state having 'a qualitative character'. So in the sentence last quoted he is asserting that there cannot be seeing that does not have some qualitative character or other. Presumably the qualitative character of Shoemaker's seeing as he 'stares' out of the window is 'things looking to one the way things now look to me'. Here we encounter the same peculiarity we noted before. Shoemaker does not *describe* the way things looked to him. Did it look gloomy out there; did the trees and plants look dusty; did the hedge look as if it needed trimming? If descriptions such as these of 'the way' things looked to him were what he had in mind, why didn't Shoemaker say so? For then he would be giving genuine examples of the 'qualitative character' of his visual experience on that occasion.

My surmise is that Shoemaker did not spell out 'the way things looked to him' because his attention was not focused on those things out there, but was instead *turned inward*. If so, the 'qualitative character' of seeing something is an 'inner picture', in the sense that an inner picture cannot be described or outwardly indicated, but can only be pointed to in an 'inner ostensive definition'. If this is so then it is no surprise that Shoemaker does not provide us with an illustration of the 'qualitative character' of seeing. This conjecture is confirmed by the way Shoemaker proceeds. He says that if a man were asked what colour a particular object looks to him he might answer that it looks blue to him (that he is 'appeared-blue-to', in the hideous jargon adopted by some philosophers). One might think here at last Shoemaker has provided a genuine example of a 'qualitative state' of a person's seeing some object: the qualitative state is that the object looks blue to him. But one would be wrong to think this. Why? Because Shoemaker believes in the possibility of 'spectrum inversion'. If spectrum inversion is possible, then, according to Shoemaker, something's looking blue to a person (his 'being appeared-blue-to') is *not* a 'qualitative state'.[26]

[25] Ibid., pp. 256–7. [26] Ibid., p. 257.

Spectrum inversion. What is 'spectrum inversion'? We can get an idea of what this is supposed to mean if we adopt the following conceit. First, we imagine that the chromatic colours are arranged on a colour wheel, like the face of a clock, with red at 12 o'clock, yellow at 3 o'clock, green at 6 o'clock, blue at 9 o'clock, and the mixed colours in between. This would be an arrangement of the spectrum of physical colours. Second, we imagine that each person has an 'inner colour wheel', a clock-like arrangement of 'colour qualia' or 'colour appearances'. This is the 'inner spectrum'. In the case of people with 'normal' colour vision we are to imagine that the inner wheel is superimposed on the outer wheel in the following way: the appearance of red on red, the appearance of yellow on yellow, and so on. To have total 'spectrum inversion' would mean that the inner wheel was dislocated from its 'normal' orientation: *no* 'inner colour appearance' would match its proper mate in the physical spectrum.

If a person were *born* with an inverted inner spectrum, no one including himself would be any the wiser, since he would learn to call red things 'red', yellow things 'yellow', etc., and would also learn to apply the phrases 'looks red', 'looks yellow', etc., as most people do. Yet none of his 'inner colour qualia' would correspond to the outer physical colours. Furthermore, a person whose 'inner colour wheel' was previously oriented in 'the normal way' might suddenly undergo 'spectrum inversion'. We can imagine this as a *rotation* of his 'inner colour wheel'. Suppose the rotation was 180 degrees: then the inner appearance of physical blue would be yellow, the inner appearance of physical yellow would be blue, etc. As Shoemaker puts it: 'if I underwent spectrum inversion tomorrow it would cease to be the case that I am normally appeared-blue-to when I see blue things, and might become the case that I am normally appeared-yellow-to on such occasions.'[27]

Because of these supposed possibilities Shoemaker concludes that 'being appeared-blue-to' is *not* 'a qualitative state'. Why does he draw this conclusion? To the best of my understanding the answer is this: Shoemaker thinks that if being-appeared-blue-to *were* a qualitative state then it would be one and the

[27] Ibid.

same qualitative state, i.e. a state with one and the same 'qualitative character'. He says: 'it is natural to make it a condition of someone's being appeared-blue-to that he be in *the* qualitative state that is, in him at that time, associated with visual stimulation by blue things.'[28] But if spectrum inversion were to occur, then the state that is 'associated with visual stimulation by blue things' would be different with different people, and might be different at different times in the life of the same person. There would be no state which was *the* qualitative state 'associated with visual stimulation by blue things'. The states that were 'associated with visual stimulation by blue things' would differ in their 'qualitative character'. There would be no 'fixed reference' for one and the same 'qualitative character'. The expression 'he is appeared-blue-to' (or, in ordinary language, 'the thing looks blue to him') would not stand for *one* specific 'qualitative character' of visual experience.

I am unable to give the notion of 'spectrum inversion' as much attention as it should receive. But as long as we are imagining things, let us imagine that there is a child who does *not* have an 'inner spectrum' but who learns the names of colours just as a normal child does, and can arrange objects according to colour, and can carry out orders such as 'Fetch me the red book', 'Paint this circle blue and that one yellow', and so on. This child arrives at the stage of discriminating and comparing colours just as the rest of us do. He also learns to distinguish, in various circumstances, the real colour of something from its apparent colour.

One may think it is *impossible* that anyone should achieve all of this if he had no inner spectrum. Would this be because it has been discovered that a person who does not have a correctly oriented inner spectrum cannot in fact discriminate colours by sight? When a motorist is given a test for colour vision is it his inner spectrum that is tested? No. The 'inner colour wheel' is a wheel that doesn't turn anything; it plays no role in determining whether someone has normal colour vision.

Inner ostensive definitions. Let us return to Shoemaker's attempt to make clear the alleged distinction between a mental

[28] Ibid. My emphasis.

state and its 'qualitative character'. He found it difficult to distinguish pain from its 'qualitative character'. So he turned to 'visual experience'—for example, seeing a thing to be blue. But he was not able to define the 'qualitative character' of a person's seeing a thing to be blue as the thing's looking blue to the person—because of the supposed possibility of spectrum inversion. How can Shoemaker provide an illustration of the 'qualitative character' of a mental state?

Shoemaker adopts the strategy of employing an ostensive definition of a 'qualitative state'. But what he offers is not a genuine ostensive definition but an 'inner' ostensive definition. He says:

> The expression 'appeared-blue-to' could, I think, have a use in which it would stand for a qualitative state. I could "fix the reference" of this expression by stipulating that it refers to . . . that qualitative state which is at the present time (April, 1974) associated in me with the seeing of blue things.[29]

Shoemaker says he is employing here the device of 'Kripkean "reference fixing"'.[30]

This device of so-called 'reference fixing' does not, however, *fix anything*. Shoemaker stipulated that the phrase 'appeared-blue-to' refers to 'that qualitative state' which, at the time of his writing, was associated in him with the seeing of blue things. But *what* qualitative state? Since Shoemaker wants us to take seriously the possibility that some people have spectrum inversion without anyone knowing it, then we must consider the possibility that Shoemaker himself has spectrum inversion. This prevents us from assuming that the qualitative state associated in him with the seeing of blue things is the state of those things looking blue to him—for they may not look blue to him. So what was the qualitative state that was associated in him, at the moment of writing, with his seeing blue things? Was it perhaps a condition of being dazzled by a reflection from one of those blue things? Was it perhaps a feeling of *déjà vu* or of frustration? How is the reader to know *what* state was associated in Shoemaker at that moment with seeing some blue

[29] Ibid. [30] Ibid., p. 258.

things, if Shoemaker doesn't tell him? Is there some one and the same state that is always associated in me with seeing blue things? Not that I know of. And even if there was, how could I, or Shoemaker, or anyone, be certain that it was the *same* state to which Shoemaker was supposedly referring?

It is important to ask why Shoemaker doesn't say *what* 'that qualitative state' was which, according to him, was 'associated in him' with the seeing of blue things? And why doesn't Nagel say *what* 'the subjective character' of seeing is? If the subjective or qualitative character of mental states is such an important topic, why don't these philosophers show concern over their failure to illustrate it? I think the answer in both cases is that by virtue of resorting to inner ostensive definitions they felt free of any doubt that they knew what *they* meant; and furthermore, since they assumed that other people too could produce for themselves inner ostensive definitions, they were pricked by no anxiety about the question of whether those other people would understand what they meant. Both Nagel and Shoemaker think that by looking inward they can pick out the subjective or qualitative character of a mental state. It is, however, nothing they can describe. The best they can say is 'It is *this*'. Obviously, Shoemaker's circumlocution ('that qualitative state which is at the present time associated in me with seeing blue things') comes to no more than 'It is *this*!'

Shoemaker's 'ostensive definition' of the phrase 'appeared-blue-to' clearly could not explain the reference of 'appeared-blue-to' to anyone other than himself. What is more, Shoemaker was not even 'fixing the reference' of 'appeared-blue-to' for himself! He was attempting to fix the reference of this expression by concentrating his attention on 'a certain impression' and bestowing on it the name 'appeared-blue-to'. If Shoemaker was actually fixing the reference of that name, or 'rigid designator', then he was undertaking to use that name on subsequent occasions to designate 'that same impression'. But since Shoemaker failed to indicate *which* impression, there could be no objective test of whether on future occasions he is right or wrong in thinking that he is using that name to designate 'that same impression'. In Wittgenstein's words, it will be a matter of 'Whatever is going to seem right to me is right.'[31] Shoemaker assumes that he could 'correctly remember'[32] 'that qualitative state'. But in the absence of any specification of 'that qualitative state', there will be no sense in saying that he remembers 'it' correctly or incorrectly. If someone declares, 'I can remember *this*', but is unable to produce any genuine ostensive definition or any description of '*this*' (and if none can be gathered from his behaviour and circumstances), then it will be meaningless to speak of his remembering or not remembering '*this*'.

Let us recall that the general point at issue is whether functionalism 'broadly construed', can account for the 'qualitative character' of mental states. Ned Block contends that 'the functional characterizations of mental states fail to capture their "qualitative" aspect'.[33] One can be in no position to agree or disagree with Block, nor to get a grip on Shoemaker's treatment of the issue, until one understands what a 'qualitative state', or the 'qualitative character' of a mental state, *is*. Now it is interesting, but not unexpected, that in order to explain this matter Block employs exactly the same device that Shoemaker does, namely, inner ostensive definition. Block says: 'One could define a qualitative state Q as a state someone has just in case he has a sensation with the same qualitative character as my current headache.'[34]

Here again we are presented with the peculiar twist of language previously encountered in Nagel and Shoemaker. Block speaks of someone's having 'a sensation with the same qualitative character as my current headache': but he does not go on to specify *what* the 'qualitative character' of his headache was. If he had said that its qualitative character was its being *piercing*, or its being *dull and throbbing*, then everyone would have a fair idea of what he meant by its 'qualitative character'. Why did not Block specify what the qualitative character of his headache was?

Like Nagel, Block feels there is something about his mental states (their 'qualitative character') that is not functionally definable. (It is what Nagel called 'consciousness' or 'the inner

[31]Wittgenstein, *PI*, §258.
[32]Shoemaker, 'Functionalism and Qualia', p. 260.
[33]Ned Block, 'Are Absent Qualia Impossible?', *Philosophical Review* 89, 1980, p. 258. [34]Ibid.

quality of experience'.) Now if Block had said that the qualitative character of his headache was its being 'dull and throbbing' and not 'piercing', then philosophers of the functionalist persuasion would surely declare that the difference between these two 'mental states' (i.e., piercing headaches and dull throbbing ones) can be defined solely in terms of differences in their respective 'inputs, outputs and successor states', i.e. differences in their 'causal roles'. It is unlikely that so nebulous a claim could be substantiated; but it is equally unlikely that it could be *refuted*. In order for Block to be absolutely secure in his contention that qualitative character cannot be captured by causal analysis, he must avoid any *description* of qualitative character. If no description of the 'qualitative character' of any mental state is provided, then of course no functional analysis of qualitative character is possible. By restricting himself to an inner ostensive definition (which comes to saying that 'qualitative character is *this*') Block makes it certain that no functional analysis of qualitative character will be forthcoming.

In discussing Nagel's position I called attention to the *regressive* feature of his metaphysical use of such phrases as 'the subjective character' or 'the inner quality' of experience. This feature is also present in Block's position. In everyday life it is a common thing for people to characterize the *quality* of their sensations: to say, for example, that their bodily pains are stabbing, burning, etc. Block, just like anyone else, could describe his sensations in these ways, but at the risk that *these* qualities of sensation will be held to be functionally definable. So Block has to say that no such quality of sensation is *what he means* by 'the qualitative character' or 'quale' of a sensation. The 'qualitative character' of his headache will not be its throbbing quality. It will be something different from but attached to ('associated with') its throbbing quality. It will be 'what it is like for me' to have this throbbing headache, where 'what it is like for me' is not specified.

What Nagel wants to mean by the 'subjective character' of seeing, or what Shoemaker wants to mean by 'that qualitative state which is at the present time associated in me with seeing blue things', or what Block wants to mean by the 'qualitative character' of his headache, is something that is left unexplained and so is not understood by anyone else; and these philosophers are deceived in thinking that they understand it themselves. They are engaged in the movement of thought that Wittgenstein characterizes in the following way: 'It is as if when I uttered the word I cast a sidelong glance at the private sensation, as it were to say to myself: I know all right what I mean by it.'[35]

3 The Causal Theory of Mind

In section 2 we considered an objection that some philosophers make against a causal analysis of psychological concepts, namely, that this analysis does not account for 'consciousness', or the 'subjective character' of experience, or the 'qualitative aspect' of some mental states. It was argued that this criticism comes to nothing, due to the apparent inability of its authors to specify *what* it is that cannot be accounted for by the causal theory of mind.

David Lewis describes a philosophical theory of mind which, according to Lewis, is held by David Armstrong as well as himself. Lewis says:

> Our view is that the concept of pain, or indeed of any other experience or mental state, is the concept of a state that occupies a certain causal role, a state with certain typical causes and effects. It is the concept of a state apt for being caused by certain stimuli and apt for causing certain behavior. Or, better, of a state apt for being caused in certain ways by stimuli plus other mental states and apt for combining with certain other mental states to jointly cause certain behavior. It is the concept of a system of states that together more or less realize the pattern of causal generalizations set forth in commonsense psychology . . . If the concept of pain is the concept of a state that occupies a certain causal role, then whatever does occupy that role is pain. If the state of having neurons hooked up in a certain way and firing in a certain pattern is the state properly apt for causing and being caused, as we materialists think, then that neural state is pain. But the concept of pain is not the concept of that neural

[35] Wittgenstein, *PI*, §274.

> state . . . The concept of pain, unlike the concept of that neural state which in fact is pain, would have applied to some different state if the relevant causal relations had been different. Pain might not have been pain.[1]

Armstrong states his view somewhat differently, although the difference does not seem to be substantial. In Armstrong's definition of the concept of a mental state his emphasis is on the role of a mental state as a *cause* of something, rather than on its being an *effect* of something. He says 'The concept of a mental state is primarily the concept of *a state of the person apt for bringing about a certain sort of behaviour*.'[2] But Armstrong adds this: 'In the case of some mental states only they are also *states of the person apt for being brought about by a certain sort of stimulus*.'[3] He regards the concept of *perception* as one of those mental concepts that are concepts of a state apt for being brought about by certain causes as well as apt for bringing about certain effects.[4] The difference between Lewis and Armstrong is that Lewis seems to hold that in regard to *every* mental state the concept of the state is definable in terms of both the typical causes and the typical effects of the state: whereas Armstrong's view is that in regard to *only some* mental states would a definition of the concept of the state include a reference to the *causes* of the state.

A number of philosophers seem to identify 'functionalism' with the causal theory of mind, taking the latter in Lewis's broad version of the notion of a 'causal role'. Ned Block, for example, says:

> Functionalism is the doctrine that pain (for example) is identical to a certain functional state, a state definable in terms of its causal relations to inputs, outputs, and other mental states. The functional state with which pain would be identified might be partially characterized in terms of its tendency to be caused by tissue damage, by its tendency to cause the desire to be rid of it, and by its tendency to produce action designed to shield the damaged part of the body from what is taken to cause it.[5]

[1] Lewis, 'Mad Pain and Martian Pain', Block vol. I, p. 218.
[2] Armstrong, *MTM*, p. 82.
[3] Ibid. [4] Ibid., p. 231.
[5] Block, 'Are Absent Qualia Impossible?', *Philosophical Review* 89, p. 257.

The causal relation. In order to get a grip on either the broader or the narrower version of the causal theory of mind, it is essential to understand what is the concept of 'cause' that plays so prominent a part in this theory. Armstrong says that the notion of 'bringing about' which appears in his definition of the concept of a mental state, 'is the "bringing about" of ordinary, efficient causality'.[6] The assumption that there is just *one* concept of 'ordinary, efficient causality', is something we must consider. Armstrong makes explicit two further assumptions about what he calls 'the nature of the causal relation'. He says:

> In the first place, I will assume that the cause and its effect are 'distinct existences', so that the existence of the cause does not logically imply the existence of the effect, or *vice versa*. In the second place, I will assume that if a sequence is a causal one, then it is a sequence that falls under some law.[7]

To put it in another familiar terminology of philosophy, Armstrong is assuming that a cause and its effect are 'logically independent' of one another, and is also assuming that a sequence of events is not a *causal* sequence unless it falls under a 'covering law'. Armstrong holds, however, that when we regard a sequence of phenomena as a causal sequence we may not know, or even have a conception of, the particular law under which the sequence might fall. Nevertheless, merely regarding the sequence as a causal one is to imply that there is a covering law for the sequence:

> In speaking of the sequence as a causal sequence, we imply that there is *some* description of the situation (not necessarily known to us) that falls under a law.[8]

Thus, Armstrong makes three important assumptions about 'the nature of the causal relation': first, that a cause and its effect are 'distinct existences'; second, that for every causal sequence there is a covering law; third, that *to speak* (and presumably *to think*) of something as causing something, is *to*

[6]Armstrong, *MTM*, p. 83. [7]Ibid. [8]Ibid., p. 84.

imply that there is some description of the situation that falls under a law.

Confusion about causality. Anscombe has rightly observed that the topic of causality is in 'a great state of confusion'.[9] On the one hand, many philosophers regard 'the nature of the causal relation', or 'our conception of causality', or 'the concept of cause', as deeply mysterious—though one may wonder why it should be more mysterious than any other concept. On the other hand, many other philosophers feel satisfied that 'the causal relation' just *is* the relation of uniform sequence between phenomena, and that a true singular causal proposition implies some true universal proposition of the form 'Whenever so-and-so then such-and-such', or at least implies that there is some such true universal proposition.

It seems to me that both sides are misled by the assumption that there is such a thing as *the* concept of cause or *the* causal relation. The fact is that the expressions 'caused', 'produced', 'brought about', 'made happen', are all used in ordinary life in a variety of different ways. Taking note of this variety helps to remove the temptation to assume that *one* formulation can capture the whole employment of the word 'cause'. Wittgenstein said of the word 'think': 'It is not to be expected of this word that it should have a unified employment; we should rather expect the opposite.'[10] In the same way we should not expect there to be a unity in the use of the word 'cause'.

Some examples of 'cause'. I am going to describe some examples of the use of the words 'cause' and 'effect', in order to show that these words apply to different relations that are *not* bound together by a common nature.

Case 1. Mr Jones creeps up stealthily behind Mrs Jones, who believes she is alone in the house, and says 'Boo!' Mrs Jones is startled. This is a case of cause and effect. An *hypothesis* as to what startled Mrs Jones would not be called for. Her startled jump was an immediate reaction to that sound behind her, when she believed she was alone. An observer would see this as

[9]Anscombe, *Intention*, p. 10.
[10]Wittgenstein, *Zettel*, §112.

an example of cause and effect and would not have to believe in the existence of a covering law. (By a 'law' I mean something that would be expressed in a true universal proposition.) A reaction such as that of Mrs Jones would frequently occur in similar situations: but it would be hopeless to try to state a law; nor do we imply that there is a law when we perceive this situation as an example of cause and effect. If Mrs Jones had known that her husband was in the house, and knew that he was prone to such pranks, perhaps she would not have made that startled jump. But one could not enumerate *all* the circumstances in which that sound would not have produced that effect, because there is no *all* to be enumerated.

Case 2. I turn a key in the lock but the bolt does not move. 'What is the cause of that?' I ask. Notice that this question does not arise in Case 1. I take the lock apart to try to discover the cause of its not working. Again, this is unlike Case 1, where no mechanism is examined. I find a piece of metal lying loose in the bottom of the lock. It occurs to me that this piece may fit between two other parts of the lock. I fix it there and reassemble the lock. When I turn the key the bolt moves. I say: 'I found the cause of the bolt's not moving: a piece of the mechanism had fallen out of place'. In this example there is wondering what the cause is; there is a conjecture; there is a testing of the conjecture. None of these features were present in Case 1.

Suppose the following objection were made: 'You did not actually *prove* that the cause of the lock's not working was the separation of that loose piece from the rest of the lock. You did not eliminate other possible causes. You tested only one hypothesis'. Now it is true that only one hypothesis was tested. But this objection would be felt to be 'academic'. Why? Because in this case the interest was solely in getting *this* lock *to work*. The interest was not in other locks, nor in conducting an 'ideal' scientific investigation. In the circumstances of this case the remark that the cause of the lock's not working had been discovered, was justified. To say that this claim was not 'proved' would be an idle use of that word.

Case 3. A and B are playing chess. Both are experienced players. To the surprise of the spectators B makes an eccentric, apparently senseless, move. When the game is finished someone asks B, 'What caused you to make that strange move?' B answers: 'I wanted to disconcert my opponent'. This answer gave B's *reason for*, his *purpose in*, making that move. This sharply differentiates Case 3 from the previous two cases. One cannot speak of the bolt's reason for not moving. One cannot even speak of Mrs Jones's reason for her startled jump: *she* did not have *a reason* for it. The unexpected sound behind her was *the reason for, the cause of it*, but was not *her* reason for jumping. In the chess example the spectators might think of various hypotheses to explain B's strange move: but when B said 'I intended to disconcert my opponent', he was not (if he remembered the move) offering an hypothesis.

There is a puzzle about the use of the word 'cause' in Case 3. The question 'What caused you to make that strange move?' would be a natural use of language in that situation. But does this imply that when B responded to the question by saying 'I wanted to disconcert my opponent', he was giving *the cause* of his move? It seems to me that the form of words the questioner used would be employed only when it was not clear to him whether B had been in a panic, or had experienced a momentary dizziness, or whether it was a failure of concentration; or whether, in contrast, the move was a shrewdly thought-out move. If the questioner had assumed it was the latter, the more natural thing for him to have said would have been '*Why* did you make that move?' instead of 'What *caused you* to make that move?' It seems to me that 'What caused you . . .?' would be said only if the questioner was uncertain whether that unusual move was carefully chosen, or whether (in contrast) the move was due to panic, or dizziness, or lack of concentration, etc. Being uncertain what *kind of explanation* there is for the move, the questioner says 'What caused it, what brought it about?' instead of '*Why* did you make that move?'

Now the puzzle is this: when the player replied, giving *his reason* for making that move, i.e. to disconcert his opponent, was he saying what *caused* him to make the move? This is a difficult question. On the one hand, might he not have replied in these words: 'What caused me to do it was that I wanted to disconcert my opponent'? On the other hand, might he not have replied in this other way: 'Nothing *caused* me to make the move; the move was deliberately calculated to upset my opponent'. It seems to me that either answer could be given. If this is right, then B's wanting to disconcert A *can* legitimately be called 'the cause' of his move, although the use of this expression in this case refers to *a different kind* of causal explanation than would be provided by the other suggested explanations.

If B can be said to be giving the cause of his move, we obtain the following consequence: it could not be said that B's making that move was 'the effect' of his wanting to disconcert his opponent! Just imagine B saying, 'My making that move was the effect of my wanting to disconcert A'! A different example may help us to see this point: if I want to climb a cliff to see what is up there, and do climb it for that reason, it cannot be said that my climbing the cliff is 'the effect' of my wanting to climb it in order to see what is up there—even though the latter can be said to be 'the cause' of my climbing the cliff.

These examples show that when the cause of a person's doing X is his reason for, his purpose in, doing X, then his doing X is not 'the effect' of his reason or purpose. Philosophers have been mistaken in assuming that *in all cases* when Y is the cause of X, X is *the effect* of Y. In the chess example, if we agree that B's wanting to disconcert his opponent can be called 'the cause' of B's making that move, we must also agree that this is not to be called a 'cause/effect sequence' (e.g. not a 'causal sequence ')— since B's making that move cannot be called 'the effect' of his wanting to disconcert A. An important consequence is that when it can be said that a person's *reason* for doing something is *the cause* of his doing it, then under that description there *cannot* be a *covering law* for the causal sequence—for there is *no* 'causal sequence'.

Case 4. Staying with the chess example, let us suppose that B's strange move *does* cause A to be disconcerted. This might be seen from an expression of astonishment on his face and from an involuntary exclamation: it might also be that after the game A admits to having been bewildered by the move. Any spectator with a fair knowledge of chess would understand why A might be bewildered by the move. A spectator's knowledge that B's making that move did cause A to be disconcerted, would in no way imply an assumption by the spectator that the situation could be described in such a way as to fall under a covering law.

Notice that in this example it can be said both that A's being disconcerted was the effect of (was brought about by, was produced by, was due to) B's surprising move; and *also* it can be said that A was disconcerted *at* or *by* B's making that move. That is, it can be said that *the object* of A's astonishment and confusion was B's making that move. As some philosophers would put it, A's reaction of astonishment and bewilderment falls under the concept of 'intentionality'. (This does not seem to me to be a happy use of language, since A's bewilderment was certainly not *intentional*.) A better way of putting the matter is to say, in Brentano's terminology, that A's reaction has 'a direction upon an object', or 'contains an object within itself'. Thus this case presents not only an undoubted example of causality, but also it is a case in which the cause and effect are *internally connected*. It is an example in which cause and effect are *not* 'distinct existences': for A's being disconcerted can only be understood as being disconcerted *at* or *by* B's strange move. It is an example of a causal sequence that *could not* fall under a *contingent* law, when the sequence is described as 'B's strange move caused A to be disconcerted.'

Going back to Case 1, where Mrs Jones was startled by the sudden sound that her husband made behind her, we can see that the same point applies. She was startled at or by that unexpected sound. It was the object of her reaction, and also it produced her reaction. It is another example of causation where cause and effect are internally related under the description, 'That sudden sound behind her caused her to be startled'. The important point for our investigation is that the 'psychological' examples of Cases 4 and 1, although they are examples of causal sequences, do not fit Armstrong's model of 'the causal relation'. This is so at least as long as the sequences are described in the natural way they are described in the examples. Perhaps Armstrong would agree, but would insist that there are other possible descriptions of those situations such that the situations would fall under contingent laws. Whether this is true will be considered later in my discussion of the concept of intention.

Case 5. In an industrial country there is a controversy over the cause or causes of rising unemployment. One party says the cause is high interest rates; another claims it is inflation; a third attributes it to the inefficiency of industry. All three parties employ statistical evidence, charts, diagrams, and historical analogies to support their respective positions. Other persons who study the evidence offered for each of the three theories are not convinced by any of it. The argument continues.

Notice how different is this use of the word 'cause' from its use in the previous examples. In this example the causal claims are linked to statistics, historical analogies, and complex argumentation—which is not so in any of the previous examples. Each party in this controversy over the cause of rising unemployment might claim to be relying on some law of industrial economy: but whether there is any such law, or whether if there were it would provide any basis for the position, is disputed by the other parties.

To summarize this brief survey of some examples of the employment of causal language: In Case 1 an unexpected sound produced a sudden start: that the one occurrence caused the other would be perceived by a spectator without inference or hypothesis, and without a comparison with other cases coming into it. In Case 2 a lock that didn't work was taken apart; an alteration was made in the mechanism, after which the lock did work; it was taken for granted that the cause of the malfunction had been discovered, although no comparison was made with other locks, and only one conjecture was tested. In Case 3 the cause of a player's surprising move is *his reason* for making the move, although his making that move cannot be said to be 'the effect' of his reason for making the move. In Case 4 the unexpected move does cause the opponent to be disconcerted, and the latter occurrence is truly the effect of that move: but here cause and effect are not 'distinct existences'. In Case 5 the dispute over the causes of increasing unemployment brings in considerations that are not present in the other examples, such as statistical comparisons and historical parallels.

The point of the survey was to illustrate how differently the terms, 'cause', 'effect', 'produce', 'bring about', are employed in the different examples. To seek a single formula for 'the link between cause and effect' or an all-embracing analysis of 'the concept of causation', is to embark on a wild-goose chase.

I have no doubt, however, that there is a use of causal language that does fit Armstrong's model. For example, the degree of flexibility of an iron bar is influenced by the temperature of the bar. The temperature and flexibility of the bar could be said to be 'logically independent' (i.e. 'distinct existences'), in the sense that they *could be measured independently* of one another. Presumably the correlation of degrees of temperature and degrees of flexibility fall under a law. If so the law could be called a *causal* law since, within certain limits, increasing temperature *causes* increasing flexibility.

In contrast, only in my Case 5 would an underlying causal law be assumed, and there it would probably be mere bluster. In some of my examples there *could not be* a contingent law underlying the phenomena *as these were described*; and any proposed 'redescription' with the aim of making those phenomena subsumable under a covering law probably would not be recognizable as a genuine description of the original phenomena.

Mental states. The causal theory of mind defines the concept of a mental state as the concept of a state that has a certain causal role. The advocates of this theory employ the term 'mental state' in an uninhibited way. Any belief, or desire, or pain, is called a 'mental state'. This is not the way this expression is used in everyday life. A twinge of pain in the shoulder cannot be called a 'mental state'—nor wanting a second cup of coffee, nor the belief that one left one's gloves in the car. In everyday language a long-term anxiety or depression is called a 'mental state': in regard to a person whom you knew to have suffered from a persisting depression, you might ask 'What is his present mental state: is he still depressed?'

Some of the causal theorists are aware that the expression 'mental state' is unsuited for many of the phenomena they want the causal theory of mind to cover. For example, Armstrong says: 'I attach no special importance to the word "state". For instance, it is not meant to rule out "process" or "event".'[11] This

[11] Armstrong, *MTM*, p. 82.

concession is not much help, since a twinge in the shoulder is no more called a 'mental process' or a 'mental event' than it is called a 'mental state'.

The causal theorists employ the term 'mental state' to cover an enormous variety of phenomena: not only physical pain, both momentary and enduring, but also knowledge, belief, hope, desire, grief, thinking, intention, understanding, remembering, etc. All of these concepts are held to be definable in terms of causal relations, causal relations that would presumably differ for each concept. Armstrong says of his version of the causal theory: 'We see the mind as an inner arena identified by its causal relations to outward act.'[12] He declares that 'causality in the mental sphere is no different from causality in the physical sphere'.[13] Causality is one and the same concept everywhere.

We noted some assumptions that Armstrong makes about 'the causal relation': a cause and its effect are always 'distinct existences'; and in speaking of a sequence as a 'causal' sequence we imply that there is *some* description of the situation that falls under a law. Perhaps not every philosopher who thinks of himself as holding a 'causal theory of mind' ascribes to these assumptions: but some do, and it is with this conception of 'the causal relation' that I will now be concerned.

Knowing the ABCs. Does Armstrong's conception of 'the causal relation' apply to concepts such as knowledge, belief, thinking, intention? Consider an example of knowing something: a friend informs me that his small child knows the ABCs. Suppose I had some doubt about this and proceeded to test the child. How would I do this? Presumably by trying to get him to recite the alphabet, asking him what letter comes after S, and so on. On the basis of his responses I would judge whether he does or doesn't know the ABCs.

According to the view of the causal theorist, the child's knowledge of the ABCs is a mental state or process that causes or tends to cause correct answers, just as the increasing temperature of an iron bar tends to make the bar more flexible. But this is an erroneous way of viewing the matter. Instruments

[12]Ibid., p. 129. [13]Ibid., p. 83.

could be employed for recording the temperature of the bar, and other instruments for determining its flexibility; and the one set of measurements could be taken independently of the other set. But what would it mean to determine whether the child knew the ABCs independently of its response to testing? It isn't as if we could detect the presence of the state of knowing the ABCs, and then further investigate whether this state produces correct or incorrect responses. After a few tests had yielded correct answers I might think that the child does know the ABCs. Suppose that further tests yielded wrong answers? Would I conclude that this is a surprising situation in which the mental state of knowing the ABCs is causing wrong answers? Certainly not. Instead, I would withdraw my previous judgement that the child knew the ABCs.

This mistaken way of thinking about the concept of knowledge suggests that it is confusing to speak of knowledge as a 'mental state'. It would not *have* to be so. One could simply employ the phrase 'the mental state of knowing the ABCs' as exactly equivalent to the phrase 'knowing the ABCs', without being misled by the connotations of the expression 'mental state'. This is possible but not likely. It is nearly inevitable that to speak of knowing, wishing, intending, etc., as 'mental states' will foster the illusion that these 'states' cause (in Armstrong's sense) their behavioural manifestations.

The strength of this temptation is apparent in Armstrong's discussion of 'knowledge by inference'. Suppose that Evans knows that *p* and later learns that if *p* then *q*. (E.g. Evans knows his dog has mange and then learns that when there is mange there is inadequate calcium in the diet.) Evans infers that *q* (that his dog's diet is inadequate in calcium). There is an ordinary use of 'because' according to which it could be said that Evans knows that *q* *because* he knows both that *p* and that if *p* then *q*. Armstrong says something remarkable here: namely, that this use of 'because' is 'simply the ordinary causal "because"'.[14] The combined mental states of knowing that *p* and knowing that if *p* then *q*, are supposed to *cause* the mental state of knowing that *q*, in the sense in which cause and effect are 'distinct existences'.

'Mental states', in the ubiquitous employment of this term by some philosophers, are nebulous items: their 'causal powers' are

[14]Armstrong, *MTM*, p. 201.

not easily discerned. Let us leave the opaque atmosphere of 'mental states', and listen to another person, Robinson, who makes the following declaration: 'I know that *p* is true and also that if *p* then *q*; but I am uncertain whether *q* is true.' Something has gone wrong here. This is evident just from the language employed by Robinson, without any investigation of the causal powers of mental states. Robinson's declaration would not stand as a counter-example to a presumed causal law. If Robinson had understood what he was supposed to be saying when he declared, 'I know both that *p* and that if *p* then *q*,' then he would have understood that it was nonsense to say he was uncertain whether *q* is true. The inferential 'because' is logical, not causal.

How could Armstrong have failed to see this? A possible explanation is that there is an ambiguity in his assumption that the cause and the effect are 'distinct existences' in the sense that the existence of the cause does not 'logically imply' the existence of the effect. On the one hand, it is true that Robinson might know that *p* and know that if *p* then *q*—but fail to infer that *q*. The more complex the values for *p* and *q*, the more likely that this might occur. This gives a sense in which the presence of the 'mental states' of knowing that *p* and knowing that if *p* then *q*, does not 'logically imply' that the 'mental state' of knowing that *q* will occur. In this sense the two sets of states are 'distinct existences'. On the other hand, if Robinson received every encouragement to make the inference, and if the values for *p* and *q* were not difficult to grasp, there would come a point, if Robinson still declined to make the inference, at which his friends would conclude that Robinson does not *understand* what he is saying, when he continues to declare that he knows that *p* and that if *p* then *q* but is uncertain whether *q*. Robinson's declaration would no longer be taken as a ground for attributing to him the 'mental states' of knowing that *p* and knowing that if *p* then *q*. In other words, the presence or absence of those 'mental states' cannot (logically speaking) be settled *independently* of the presence or absence of the 'mental state' of knowing that *q*. This gives a clear sense in which these 'mental states' are *not* 'distinct existences', as are the temperature and the flexibility of an iron bar.

Genuine duration. Another harmful result of calling the instantiations of all psychological concepts 'mental states', is that it produces a deceptive appearance of homogeneity in these concepts. A useful method for discerning differences is to reflect on how temporal *duration* applies in a variety of examples. In *Zettel* 45 Wittgenstein introduces the term 'genuine duration' (*echte Dauer*). He says that an *intention* 'is not a state of consciousness. It does not have genuine duration.' This could be misleading, since an intention (the having of an intention) *does* have duration: for we can ask, 'How long has he had that intention?' But there is no need to be misled: Wittgenstein is distinguishing between *different concepts* of duration. He is using the term 'genuine duration' as a technical term to stand for one of those concepts. He is not implying that other concepts of duration are sham or unreal.

What is genuine duration? In various passages in *Zettel* there are indications of how this term is to be applied. For example: 'Where there is genuine duration one can tell someone: "Pay attention and give me a signal when the picture, the thing you are experiencing, the noise, etc., alters".'[15] 'Think of this language-game: Determine how long an impression lasts by means of a stop-watch. The duration of knowledge, ability, understanding, could not be determined in this way.'[16] '"I have the intention of going away tomorrow".—When have you that intention? The whole time; or intermittently?'[17] 'Consider what it would mean "to have an intention intermittently". It would mean something like: to have the intention, to abandon it, to resume it, and so on.'[18]

To try to define genuine duration would be difficult and probably useless. What are needed are examples: examples of things that have duration but not genuine duration, and examples of things that have genuine duration. The concept of genuine duration is connected with the concepts of keeping something under continuous observation, or of focusing one's attention on something, or of being on the alert to note or report or signal, changes in the quality or state of something. For example, a ship's lookout has a submarine in sight through his binoculars, and is to keep reporting whether he still sees it, or

[15]Wittgenstein, *Zettel*, §281. [16]Ibid., §82.
[17]Ibid., §46. [18]Ibid., §47.

whether it has disappeared from his view: ('I see it', 'I still see it', 'Now I don't see it'.) He would be indicating the duration of his visual perception of the submarine: this would be genuine duration. At a children's party there is a competition to determine which of them can stand on one foot for the longest time. The judge uses a stop-watch. The duration of standing on one foot in this game would be genuine duration. In another contest, each contestant throws a ball. The one wins whose ball remains in motion for the longest time: genuine duration. In a laboratory a careful check is maintained on whether the temperature of a liquid goes above or below certain points on a gauge. The duration of the temperature between those limits would be genuine duration.

Let us consider some other examples. A couple have been married for ten years. Is the duration of their marriage genuine duration? No. It would make no sense to keep them under constant surveillance to ascertain whether they are still married. Nor would it make sense to keep asking them, every few minutes, day and night, 'Are you still married?'—as it would make sense to keep asking the lookout, 'Do you still see the sub?'

A man has been a British citizen for several years, is able to read French, likes to go fishing on weekends, and hopes to write a novel. Each of these things has been true of him for several years, but none of them has genuine duration. It would make no sense to require him to pay close attention and to report from moment to moment whether he is still a British citizen, hopes to write a novel, likes to fish, etc.

If a man were suffering from periodic spasms of pain, he could be asked to signal the coming and going of the spasms by pressing a button that turns a light on and off. It might also be possible for his doctor to measure the duration of each spasm, not by relying on any such signal, but by closely observing the patient's facial contortions and the tightening and relaxing of his limbs and muscles. The duration of the spasms determined in these ways would be genuine duration.

If a man in a pub flew into a rage because of an insult, it would scarcely make sense to keep *asking him* from moment to moment whether he is still in a rage. But an observer could determine the duration of this violent emotion from the opening outburst to his calming down, by noting the expressions of rage in his countenance, gestures, and exclamations. The duration of the period of rage that would be determined in this way would be genuine duration.

A child outdoors is frightened by a dog and runs for the house in a panic of fear. Once he is safely inside his fear subsides. Here again it would be absurd to expect the child to keep reporting during the panic whether he is still afraid: but an observer could time the duration of the panic from its eruption to its cessation.

One could observe a person's surge of hope (that he will be rescued) followed by the dying of hope. One could observe that a person was gloomy throughout dinner, but later cheered up. This observation of the duration of both the period of hope and the period of gloom would be the observation of genuine duration.

In summary, the concept of genuine duration applies to some physical phenomena, such as the state of motion or rest of an object, the constant or changing temperature of a liquid or of the pressure of a gas or of the electric charge of a substance, and so on. Genuine duration also applies to some psychological phenomena, such as sense-perception, bodily sensations of heat and cold, dizziness, pain; also to after-images. In these cases the subject could signal the beginning and ending of the sensations or images, or the changes in them. Genuine duration also applies to the rise and decline of an emotion or mood, in cases where the subject could not be expected to give signals or reports, but where a spectator could perceive the duration of the emotion or mood via the subject's expressive behaviour.

There are important differences between some of these psychological concepts. Knowledge or belief would enter into some of them but not others: e.g. the man in a rage believes he has been insulted; the man who had a surge of hope believed for a moment he would be rescued: but a man suffering the violent pain of a leg-cramp has no relevant belief—certainly no 'belief' that he is in pain. Genuine duration applies despite these differences in the concepts.

It should be noted that it is impossible to say whether genuine duration applies to a concept until one knows something about the context in which the concept is employed. A man may have been hoping for the past two years that he would be promoted to be the manager of his firm: in this use of the word 'hope' genuine duration does not apply: but it does apply to the situation where a man in peril of his life has a momentary 'surge of hope' that he will be rescued. Genuine duration pertains to some cases of *hoping* and not to others: the same is true of *belief*, and so on.

States of consciousness and dispositions. The too generous employment of the term 'mental state' by philosophers conceals important differences between psychological concepts. I am thinking particularly of the fact that genuine duration applies to some of these concepts but not to others. The following remarks of Wittgenstein may be of help here:

> I want to talk about a 'state of consciousness', and to use this expression to refer to the seeing of a certain picture, the hearing of a tone, a sensation of pain or of taste, etc. I want to say that believing, understanding, knowing, intending, and others, are not states of consciousness. If for the moment I call these latter 'dispositions', then an important difference between dispositions and states of consciousness consists in the fact that a disposition is not interrupted by a break in consciousness or a shift in attention.[19]

> Intent, intention, is neither an emotion, a mood, nor yet a sensation or image. It is not a state of consciousness. It does not have genuine duration. Intention can be called a mental disposition. This term is misleading inasmuch as one does not perceive such a disposition within himself as a matter of experience. The *inclination* towards jealousy, on the other hand, is a disposition in the true sense. Experience teaches me that I have it.[20]

The distinction Wittgenstein wishes to draw with the help of the expressions 'state of consciousness' and 'disposition' is important. These expressions are not commonly employed to mark out the distinction he has in mind: Wittgenstein is simply choosing them, as terms of art, for his purpose. I think the term 'disposition' is not a happy choice. Knowledge, for example, is

[19] Wittgenstein, *Remarks on the Philosophy of Psychology*, vol. II, edited by G. H. von Wright and Heikki Nyman, translated by C. G. Luckhardt and M. A. E. Aue, Blackwell, 1980, §45.

[20] Ibid., §178.

not, at least in many uses of the word, a disposition. If Mary knows how to do the laundry that does not imply that she has a disposition to do it, or even to teach someone how to do it: what she does have is the *ability* ('know-how') to do the laundry. In contrast, the inclination to be jealous is, as Wittgenstein notes, a disposition.

On the other hand, the expression 'state of consciousness' is a happy choice, for this reason: if a person is undergoing any of the psychological phenomena that have genuine duration then the person is *conscious*: in contrast, if a person believes, knows, or understands something, it does not follow that the person is conscious. True statements can be made about a person's beliefs, knowledge, opinions, intentions, even if the person has been made temporarily unconscious by a general anaesthetic or a blow on the head. But if someone is having a sensation of smell or taste, is seeing or hearing something, is having an image or feeling pain, is bursting with rage, or feeling sudden dismay or sudden hope, the person is conscious. The things that imply consciousness are not always *states*. Having a sudden thought, or suddenly remembering an engagement, are mental occurrences or events, rather than states.

Intention. Today you decided to go to Spain next summer, and a month later a friend asks whether you still have that intention. The answer 'No' would mean that you had given up the intention; the answer 'Yes' would mean that you still have it. But your continuing to have that intention is nothing like experiencing a surge of anger or hope, or a spasm of pain, or the occurring and then fading of an after-image. Your having that intention is not a sensation, image, or feeling; nor is it a thought. There is no facial expression of intention, as there might be one of determination. There is no cry of intention, as there may be a cry of hope or dismay or rage. There is no behavioural expression of your having that intention which someone could observe throughout the period of your having it. Your intention does not have an experiental content, something coming and going, emerging and fading—such that you could report it in words like, 'Now it's there; now it's gone; now it's there again'. You have had that intention for a month, although during that time you have been asleep for many hours, and might even have been unconscious for a time. Having this intention is not a state of consciousness: it does not have genuine duration. Nor is having this intention an occurrence, although your decision, your forming that intention, was an occurrence.

Armstrong says the following about the concept of having an intention:

> The having of purposes and intentions are not events, or processes, but are *states* of our mind. They are states with causal powers: powers to initiate and sustain trains of physical or mental activity.[21]

> Suppose we consider the stretch of time from the moment a purpose or intention is formed up to the moment it is fulfilled or abandoned. At each moment we have the purpose or intention.[22]

A month ago you decided to go to Spain in the coming summer and you still have that intention. What would it mean to say that 'at each moment' during the past month you have had that intention? *Not* that at every moment, day and night, asleep and awake, you were monitoring that intention and noting that it was still there. It is unclear what would be meant by saying that 'at each moment' during the month you have had that intention. It might merely mean that at no time during the month did you change your mind about going to Spain, nor waver in your thinking about the matter. Let us assume that *that* is how it was: you did not give up your intention even temporarily, nor did you waver in your thinking about it. But this description is merely *negative*: it says that a certain occurrence did *not* take place. There is no implication that something was continuously present ('a state with causal powers') during the entire month. But it is this implication that Armstrong wants to draw from the description of the case.

It is possible to think of a special context in which the phrase 'at each moment' or 'at every moment' might be meaningfully applied to having an intention. Suppose a climber is working

[21]Armstrong, *MTM*, p. 136.
[22]Ibid.

his way up the side of a mountain with the intention of reaching the summit. He is caught in a severe snowstorm, is suffering from the bitter cold and from exhaustion: but he won't give up; he struggles on. It might be said of him later on: 'Despite his exhaustion and suffering and the terrible odds against him, he kept at every moment to his intention to reach the summit'. This climber was continuously screwing up his resolve. Such a thing could not be said of someone who is planning to take his holiday in Spain, or of someone who is walking down the street with the intention of buying a newspaper. Even in the case of the dauntless climber who 'at every moment kept to his resolve', could it be said that 'at each moment he was aware of his intention', as at each moment he was aware of his exhaustion and pain? If this were said, what would it mean? Probably it would be no more than a fancy way of saying that at no time did he consider turning back.

When a philosopher holds that intending to buy a newspaper, knowing the knight's move in chess, wanting to learn Italian, hoping for sunny weather, believing that Finnish and Swedish are the same language, etc., are 'states' of a person, it is impossible to agree or disagree with him until one understands what he is trying to say. If in calling them 'states' he means to assert that intending, knowing, etc., have genuine duration, then he is wrong. Calling them 'states', just in itself, means nothing at all. This use of the term 'state' could be purely redundant: 'He is in the state of intending to do X' may mean nothing more than 'He intends to do X'.

Knowing, believing, intending, etc., are not events or processes. Making a decision at a certain moment, would be an event. Cleaning the kitchen, preparing dinner, playing a tune on the guitar, are processes. It is tempting to call knowing, believing, intending, etc., 'dispositions'. But this is not how that word is used. Being normally good-humoured, having a tendency to be irritable—these are dispositions. But not, knowing the knight's move or intending to get a newspaper. There does not seem to be any general term of ordinary language that covers all of these concepts. Knowing the knight's move is an ability, i.e. the ability to move the piece correctly in play and to show someone how it moves. Intending to buy a newspaper is not an ability, nor is hoping for hot weather. It is probably exasperating for a philosophical theorist not to have available a term that embraces all of the instantiations of psychological concepts, for this hinders his desire to speak with generality. His temptation is to take over a term such as 'mental state', which does have a use in ordinary language although a very limited one. But stretching this term far beyond its normal application, obscures the meaning of his philosophical theory right from the start.

Materialism. I suspect that Armstrong's predilection for the term 'state' is partly due to the materialist inspiration of his causal theory of mind. According to Armstrong's 'central-state theory' the 'inner mental states' are identical with 'physico-chemical states of the brain'.[23] Intending to go to the opera is one state of the brain, hoping for an inheritance is another state of the brain, knowing how to square numbers is still another, etc. A physico-chemical state of the brain is something that could be monitored with instruments. Changes in chemical balance or quantity of electrical charge in some area of the brain, or changes in the firing pattern of a cluster of neurons, could be monitored continuously and recorded by instruments. This means that brain-states and processes do have genuine duration. Therefore (it might be argued) *if* intending, hoping, knowing, etc., *are* states of the brain then they too have genuine duration.

Of course this reasoning can be stood on its head. Since intending, hoping, knowing, remembering, etc., do *not* have genuine duration, and physico-chemical brain-states *do* have it, then intending, knowing, etc., are *not* brain-states. One could say that the 'central-state theory' is false on logical grounds. But perhaps it would be better to say that this theory is not intelligible. I find it impossible to understand what could be meant by saying that something, A, that has genuine duration, and something, B, that does not, are really one and the same thing.

The causal theory of mental states. Let us return to the causal theory of expecting, intending, knowing, understanding, etc.

[23]Ibid., p. 91.

The view is that these are 'mental states' and that the differences between them consist in differences between their typical causes and their typical effects. Now clearly there is a logical or conceptual connection between, for example, knowing the knight's move, and the actual practice of moving the piece correctly in play and/or giving correct explanations or demonstrations of its move. By speaking of a 'conceptual' connection I mean that it could not be said that a person 'knows the knight's move' unless his movements of the piece were mostly in accord with the rule for the knight's move. How then can it be maintained that knowing the knight's move is a state that *causes* (in Armstrong's sense of 'causes') the behaviour of correct play with the piece? For on Armstrong's view, 'the causal relation' is a *contingent* relation between cause and effect, not a conceptual or logical one.

Armstrong is well aware of this difficulty, and in his treatment of the concept of *intending* he makes an admirably straightforward attempt to overcome it. First of all, he asks whether 'the connection between the intention and the occurrence of the thing intended' is purely contingent'.[24] He goes on to say, 'There seems to be some logical bond between intention and the occurrence of the thing intended that there cannot be between ordinary cause and effect.'[25]

It isn't that there 'seems' to be a logical bond between intending to do X and doing X: there really is one. This makes the concept of intending an especially suitable test case for Armstrong's causal theory of mind. But it would be well to clarify the nature of this 'logical bond'. Roughly speaking, we do not ascribe to a person the intention to do something, X, unless we expect him to do X, or at least attempt to do X. Of course this is not so if we believe that doing X is impossible, or impossible for him. Also it can happen that after a person has announced his intention to do X, he gives up that intention for some reason or other. Even if he keeps to his intention various unforeseen circumstances might prevent him from doing X. Suppose Jenkins announced to his wife that he was going to repair the roof during his summer holiday: but when the time came he was ill, or the weather was bad, or he found that the

[24]Ibid., p. 133. [25]Ibid., p. 134.

cost of materials was prohibitive, or other matters turned up that required his full attention, or he simply forgot about repairing the roof, and so on. Such unanticipated events and facts might be called 'countervailing circumstances'. The presence of countervailing circumstances would justify one in saying that Jenkins did intend to repair the roof even though he did not fulfill the intention. On the other hand, if there did not appear to be any countervailing circumstances, and despite reminders from his wife Jenkins did not even make an attempt to repair the roof during his holiday but went fishing instead, one would be justified in thinking that Jenkins had not 'seriously' intended to repair the roof during his holiday. This brings out the nature of the logical connection between *intending* to do something and *doing* it. If doing it is well within the person's powers, and if he has not given up the intention for some reason or other, and if he has not forgotten his intention, and if no countervailing circumstances have arisen, and if he offers no satisfactory explanation for not fulfilling that intention, and so on—then if he doesn't do the thing, we would conclude that he does not really intend to do it. This way of judging the matter is *required by the concept of intention*. It is a judgement based on a concept, not on experience of life. Whether a person does or does not do what he said he intended to do, is *not indifferent* to whether it is right to regard him as having had that intention. There are people who frequently declare, in all earnestness and sincerity, that they intend to go on a strict diet, or to give up smoking or drinking or drugs, but who never do so. After a while one no longer believes those declarations. The logical bond, the conceptual connection between intending and doing, is a loose one; nevertheless it is strong enough to rule out the possibility of there being a merely contingent connection between intending and doing.

The double description theory. When Armstrong says that 'there seems to be some logical bond between intention and the occurrence of the thing intended that there cannot be between ordinary cause and effect,' I think he means there actually *is* a logical bond, of the sort I have tried to describe, between having the intention to do X and doing X. Armstrong holds, however, that this is so *only* if the intention is referred to by the description 'the intention to do X'. If the intention could be referred to by a different sort of description, a 'neutral' one, then under that description the 'mental state' of having the intention would be *contingently* connected with the behaviour of doing or trying to do X. Armstrong takes the position that whether a mental state, M, is contingently or non-contingently related to some behaviour, B, is wholly a matter of how M and B are described. The contingency or non-contingency of the relation is *relative* to the descriptions of M and B. This position seems to me to be sound in itself. The question is whether it can be applied to intentions.

Armstrong proposes an analogy between the concept of intention and the concept of *being brittle*. He says that 'it is no mere contingent fact that brittle things regularly break'.[26] It is true that to call a glass 'brittle' implies that it is likely to break or shatter if dropped or struck. A brittle thing is 'apt for breaking'. There is a logical link between the concepts of being brittle and of tending to break easily. Fair enough. But, says Armstrong, there is another way of characterizing the brittle glass:

> Although speaking of brittleness involves a reference to possible breaking, the state [of being brittle] has an intrinsic nature of its own (which we may or may not know), and this intrinsic nature can be characterized independently of its effect. And it is a mere contingent fact that, in suitable circumstances, things with this nature break. Now may not the relation of the intention to the occurrences of the thing intended stand in much the same relation as brittleness stands to actual subsequent breaking? And, if so, intentions may still be causes of the occurrence of the thing intended.[27]

Armstrong is proposing what might be called a 'double description theory'. The glass might be characterized simply as being brittle, in which case the relation between its being brittle and having the tendency to break easily, (its being 'apt for breaking'), would be non-contingent. Or the glass might be characterized in terms of its 'intrinsic nature', which might be its molecular composition; and the relation between this

[26]Armstrong, *MTM*, p. 134.
[27]Ibid.

physical constitution and the tendency to break would be contingent. Similarly, my having a certain intention can be described in two different ways. (Presumably the same holds for other 'mental states'.) The first sort of description of my having that intention, and the only description that is available to me 'by introspection',[28] is simply the description, 'I intend to do X', i.e. a description of my intention in terms of its *object*, i.e. *what* I intend to do. As Armstrong puts it, my intention is a 'mental cause':

> My direct awareness of this mental cause is simply an awareness of the sort of effect it is apt for bringing about. It is this that prevents us from being able to characterize the mental cause, except in terms of the effect it is apt for bringing about, and which gives the appearance of a quasi-necessary connection between cause and effect.[29]

By analogy, however, with the brittleness of the glass, an intention has an 'intrinsic nature' which could in principle be characterized independently of its effect. Let us call this supposedly possible way of describing an intention an 'intrinsic description' of the intention. According to Armstrong, the relation between an intention, under its intrinsic description, and the behaviour of bringing about or of trying to bring about the intended state of affairs, is a purely contingent relation. He says:

> The peculiar 'transparency' of such mental states as the having of intentions, our inability to characterize them except in terms of the state of affairs that would fulfill them, is thus explained without having to give up the view that purposive activity is the effect of a mental cause, and an ordinary contingent cause at that.[30]

Reductio ad absurdum. Armstrong's proposal is certainly ingenious, and at first sight it is attractive. My intention (e.g. to paint the bathroom) is subject to two modes of description. According to the only familiar mode of description it is

[28] Ibid. [29] Ibid., pp. 134–5. [30] Ibid., p. 135.

characterized simply as the intention to paint the bathroom. Let us call this its 'intentional description'. Under its intentional description there is a 'logical bond' between this intention and the behaviour of painting the bathroom: the relationship is non-contingent. But this same intention is supposed to have an 'intrinsic nature' that could, in theory, be described. Under the intrinsic description of the intention the relation between the intention and the behaviour would be contingent. Thus the intention and the behaviour are related *both* contingently and non-contingently.

Intentions do have intentional descriptions. But why should we think they might have intrinsic descriptions? Armstrong believes in the possibility of intrinsic descriptions of intention, and of other 'mental states', because he thinks that mental states are states of the brain. My intention to paint the bathroom just *is* a particular state of my brain.

Previously I argued against the identification of an intention (or the having of an intention) with any physico-chemical or neurological state of the brain. The argument was that any such brain-state would have *genuine duration*, whereas a person's intending to do so-and-so does not have genuine duration. Something, A, that has genuine duration, and something, B, that does not, cannot be said to be identical. My present argument will be different. It will take the form of *reductio ad absurdum*. I will try to show that a logically unacceptable consequence follows from the conception that an intention, under its intrinsic description, is related contingently to the behaviour of carrying out the intention.

Armstrong's view is that there is a state of my brain that is identical with my intention to paint the bathroom. This brain-state, if we knew enough, could be referred to by a physico-chemical or neurological description, which would imply nothing whatever about any behaviour of mine in which it would tend to issue or be apt for bringing about. The brain-state that is supposedly identical with my intention could, we may suppose, be specified quite accurately. It might, for example, be specified as a cluster of individual neurons that fire in a certain pattern. Under its neural description this brain-state, call it 'S', is supposed to be logically independent of the behaviour of painting the bathroom. S is supposed to *cause* that behaviour: but this cause, under its intrinsic neural description, has no kind of conceptual tie with that effect, under the latter's description, 'painting the bathroom'. So described, cause and effect are 'distinct existences'.

It is also part of Armstrong's view that there is some description of the causal sequence according to which the sequence falls under a *law*. I take him to mean that the law would be *contingent*: if it were not, the relation of cause to effect would not be contingent. Presumably the law would be of the form, 'Whenever X then Y'. The law would be a description of a universal regularity.

It follows from Armstrong's view that the neural state, S, might (in a world with different causal laws) have caused entirely different behaviour. Instead of causing me to paint the bathroom it might have caused me to write a poem. Now we have to bear in mind that S is supposed *to be* my intention to paint the bathroom. If this were true it would follow that my intention to paint the bathroom *might have been* my intention to write a poem. This is easily seen to be a consequence of the causal theory of mind of Armstrong (and also of David Lewis): for according to this theory the concept of a mental state is the concept of a state that 'occupies a certain causal role'. If neural state S (which we are assuming to be the intention to paint the bathroom) *might* have been 'apt for causing' poem-writing behaviour, then it follows that my intention to paint the bathroom *might* have been my intention to write a poem.

We can see that this is an absurd consequence if we reflect on how we distinguish and individuate intentions. Whether an intention A, and an intention B, are the *same* intention, or are *different* intentions, depends solely on *what* is intended—on the *object* of intention. The object of an intention (what one intends to do or bring about) is essential to the intention's being the intention it is. Same object, same intention: different object, different intention. This is also how we individuate expectations: if what you expect is different from what I expect then we have different expectations. This is also how we individuate belief, hope, fear, etc.

Let us be clear that the implication of Armstrong's theory is *not* just that *instead* of intending to paint the bathroom I might have intended to write a poem. No. The implication is that *the one* intention *might* have been *the other* intention. Of course another implication is that if the causal laws relating 'mental states' to behaviour had been still different, then my intention would not even have been *an intention*, but might have been a feeling of remorse, or a liking for ice cream, or perhaps not even a 'mental state' at all. These absurd outcomes result from the combination of (1) Armstrong's materialistic mind–brain identity theory and (2) his causal theory. The first theory asserts that a specific brain-state may be identical with a particular mental state; the second theory implies that if that specific brain-state had had a different causal role, then it would have been a different mental state, or perhaps no mental state at all. Thus, we obtain the nonsense that a certain intention *might* have been some other intention, or something other than an intention.

Causality in mind. The argument from *genuine duration* is a refutation of the mind–brain identity thesis, not of the causal theory. The *reductio ad absurdum* argument is a refutation of *the conjunction* of the identity thesis and the causal theory. Neither of these goes against the causal theory of mind, taken by itself. What can be said about *it*?

Well, we have already said quite a lot against it in discussing those psychological concepts that are not concepts of either states, events, or processes, of consciousness (in the sense that their instantiations in a person at a certain time do not imply that the person is conscious at that time). *Intending* is one of those concepts. In our discussion of it we noted that although one can say of a person that certain actions of his are *due to* or *because of* his intending to do X, this 'causal' relation invoked here is not the kind of causal relation that is required by Armstrong's 'causal theory of mind', since it is a 'logical bond' and not a contingent relation.

But there are many psychological concepts that are concepts of states, processes, or events of consciousness. A moment of panic, a burst of pain, a sudden decision, a thought flashing through one's mind, suddenly remembering an appointment —these all imply that the person is conscious. How does Armstrong's causal theory fare in regard to such concepts?

Let us compare the concept of a moment of panic with the concept of a sudden burst of physical pain. Suppose that in social conversation I make a remark intended to be humorous, but immediately realize that the remark may be offensive to one of my guests. I feel a moment of panic, and then apologize for the remark. Here one can speak of 'cause' and 'effect'. The feeling of panic was caused by my realization that my inept remark may have been offensive. This realization, and my dismay, caused me to apologize. Suppose that the next day I related this incident to someone, who responded as follows: 'I know exactly what your panic was like. Yesterday, when crossing the street, I suddenly saw a car bearing down on me. I felt just the same panic that you did'.

This would be a ludicrous comment. Why? Because *the causes* of panic were so different in the two cases that this other person cannot be said to have felt *the same panic* that I did. One could put the point in this way: a feeling of panic and its cause are 'internally' related. What is called 'the cause' of the feeling of panic can be said to be what the panic is *about*. The connection between a feeling of panic and its cause, what the panic is about, is a conceptual connection and not a contingent relation.

Consider now my having a burst of physical pain. I am ignorant of the cause of the pain, but I can describe the pain. The description might mention its *location* ('in my chest'); its *quality* ('a burning sensation'); its *intensity* ('very acute'); its *duration* ('lasted only for a moment'). I do not know the cause of the pain but this does not prevent me from describing it. Here we have a sharp contrast with the example of a feeling of panic. I could not describe that feeling of panic except by mentioning what caused it, i.e. what it was about. But I could describe the pain without saying anything about its cause. In the case of the feeling of panic the cause of it belongs to the concept of it; in the case of the burst of physical pain this is not so.

There is, however, a possible source of confusion here. If the bursts of pain in my chest occurred frequently, and I went for a physical examination, the cause of this pain might be discovered, e.g. a lesion in the heart. The doctor, and subsequently I myself, might describe the pain as 'angina pain', thereby making a reference to its cause 'internal' to the description of *that* pain. A pain of 'intrinsically' the same description (in terms of location, quality, intensity, duration) but with a different cause (e.g. indigestion) would be called *a different pain*. Thus, two modes of description are possible for a physical sensation, one including a reference to its causation, the other not. The reference to its causation is not a reference to what it is *about*. In contrast, what caused the feeling of panic is what it was about. The reference to the situation that caused the panic is, right from the start, 'internal' to the description of the panic—which is not so of the cause of the burst of pain.

It is a curious fact (though probably no coincidence) that the most frequent example of a 'mental state' occurring in the writings of the causal theorists is the example of bodily pain. Their conception of 'the causal relation', namely, that cause and effect are describable independently of one another, is actually satisfied by a bodily pain such as the one just described: the burst of pain and its cause, really are 'distinct existences'. (It is irrelevant that once the cause is known, a reference to this cause might, more or less gradually, become a part of the concept of that (type of) pain.) This particular feature of the causal theory of mind is true of this example—which, ironically, is *not* a mental state. But many of the psychological concepts the causal theorists want their theory to cover (e.g. the so-called 'dispositions', such as knowing, intending, expecting; or momentary conscious occurrences, such as a feeling of panic, or being startled by an unexpected sound, etc.; or, still different, the chess player's bewilderment caused by his opponent's strange move—do *not* satisfy the causal theorist's assumption that cause and effect are 'distinct existences', i.e. that cause and effect are related externally, not internally. Therefore to be fair we must conclude, not that the causal theory of mind is completely false, but that it is mostly false.

Functionalism again. Recent philosophy of mind has been dominated in some quarters by functionalism. This, it will be recalled, is the theory that a mental state is a functional state, and that a functional state is defined by its causal role. We have been concentrating on the particular view that those functional states which are mental states, are states of the central nervous system. But among those philosophers who hold that mental states are functional states, there are some who either deny or are in doubt about the claimed identity of mental states with neural states. One reason they offer is that different species of animals, species with different 'neuroanatomical structures', might have some of the same mental states. Richard Boyd says: 'It is highly unlikely that there is any quite specific neurophysiological state common to, for example, all animals that are in pain.'[31] The imagined identity of pain with the firing of C-fibres (much favoured by philosophers) cannot hold if there are animals in pain who have no C-fibres.

Another reason advanced for doubting that there are any identities between mental phenomena and neural phenomena is the belief that mental phenomena can be instantiated in *machines*. According to Boyd,

> It is logically possible—indeed, even physically possible—for these phenomena to be realized by entirely inorganic mechanical computers and, thus, that they can be realized by systems that possess no *physiological* definition whatsoever.[32]

Fodor subscribes to this same reason for doubting that 'there are neurological kinds coextensive with psychological kinds':

> For it seems increasingly likely that there are nomologically possible systems other than organisms (viz., automata) which satisfy the kind predicates of psychology but which satisfy no neurological predicates at all.[33]

Thus, it is frequently claimed by functionalists that the same mental state can be physically 'realized' in a variety of different ways. Shoemaker puts their point in this way: 'The fact that functional states are "multiply realizable" implies that a functional state cannot be identical to any particular physical realization of it . . .'[34]

[31] Richard Boyd, 'Materialism without Reductionism: What Physicalism Does Not Entail', in Block vol. I, p. 91.

[32] Ibid., p. 93.

[33] Jerry A. Fodor, 'Special Sciences, or The Disunity of Science as a Working Hypothesis', in Block vol. I, p. 125.

[34] Sydney Shoemaker, 'Some Varieties of Functionalism', *Philosophical Topics* 12, 1981, pp. 97–8.

This would still leave open the possibility that a functional state (and therefore a mental state) must be 'realized' in some physical state or other. One might suppose that this would be a minimum requirement for the materialist (physicalist) allegiance of the functionalist view. But some functionalists don't think so. Boyd says:

> There seems to be no barrier to the functionalist materialist's asserting that any particular actual world mental event, state, or process could be—in some other possible world—nonphysically realized.[35]

So this version of functionalism envisages the possibility that mental states might have a non-physical realization! Are we to gather from this that these functionalists no longer regard themselves as physicalists? By no means. For what they hold is that in our world mental states are 'realized' solely in physical states, but that 'in some other possible world' they might be 'realized' non-physically.

Apparently we are to conceive that in human beings the mental state of pain may be 'realized' by the firing of C-fibres, in chipmunks by the firing of G-fibres, in rats by the firing of H-fibres, in a mechanical computer by the motion of metal parts, in an electronic computer by occurrences in electrical circuits and silicon chips, and in a non-material soul by the agitation of spiritual elements.

Realization. Although I have disagreed both with Armstrong's causal theory of mind and his mind–brain identity thesis, his presentation is lucid: one can get a grip on his view. The same cannot be said for those functionalists who subscribe to the causal theory but wish to avoid the identity thesis. For one thing, their use of the terms 'realize' and 'realization' is baffling. In ordinary language we speak of a plan, a hope, a desire, an intention, as having been realized. ('His long-standing desire to see the Taj Mahal was at last realized'; 'Her hope to become chief editor was realized'.) In this use 'realized' comes to the same as 'fulfilled', 'achieved', 'satisfied'. The realization of a desire is the occurrence of what was desired. The realization of

[35] Boyd, 'Materialism without Reductionism', p. 101.

a man's desire to see the Taj Mahal would be *his seeing the Taj Mahal*. To say that a woman's hope to become chief editor was realized, would mean that she actually became chief editor.

How different is the use of 'realized' by the functionalists! According to them, a man's desire to see the Taj Mahal would be realized even if he never got to see the Taj Mahal; it would be realized by one of his neural states. The woman's hope to become chief editor is *already* realized—in her brain!

We have a right to expect from the functionalists a coherent account of what they mean by a 'realization'. By saying that a mental state is realized by or in a neural state, do they mean that the latter produces, or is correlated with, the mental state? Apparently not: for either view would smack of mind–body dualism, and the functionalists want to be materialists. Could they mean that to speak of a mental state as being 'realized in' a brain-state, is to say that the mental state just *is* the brain-state that realizes it? This is Armstrong's view, and it is straightforward. Among the functionalists there is much waffling over whether or not they want to accept this identity thesis. For example, K. V. Wilkes, in referring to some imagined neurophysiological process, says:

> For such a process to happen *is* for the organism to experience pain. By saying this, the physicalist has become a monist. But this 'is' should not be thought to import a thesis of a strong or Leibnizian identity of pains and brain processes . . . Certainly the physicalist wants to argue for monism of *some* kind . . . but his best bet is to avoid the entire identity dispute.[36]

So how is monism without '*strong*' identity to be achieved? According to Wilkes, by recourse to 'eliminative materialism'. What is *that*? It is the view that, for example,

> Our entire sensation discourse, at least in its descriptive and explanatory roles, could in principle be replaced by descriptions and explanations that referred solely to brain processes.[37]

How would this 'replacement' occur? Well, in this way: in the

[36]K. V. Wilkes, *Physicalism*, Routledge & Kegan Paul, 1978, p. 101.
[37]Ibid.

future children would be taught to walk on tiptoes when told that their father's 'C-fibres are firing', instead of, as now, when told that he 'has a bad headache'.[38] What a marvellous solution! One expression is replaced by another. Would this *eliminate* sensation discourse? Not at all. For the new expression would not *take the place* of the familiar expression unless it were given the *same use*. For example, the first-person expression, 'My C-fibres are firing', would have to be used as an immediate expression of sensation, not as a hypothesis about what is going on in the speaker's brain, nor as a report of an observation made with the help of instruments. But then nothing would have changed except a bit of terminology.

Functionalist writings exhibit considerable confusion over the question of whether a mental state is *identical* with that which 'realizes' it. Boyd, in one passage, speaks in favour of 'rejecting the identity of a token event, state, or process with its actual realization'.[39] But in a subsequent passage Boyd says that 'mental states are *in fact* central-nervous-system states but . . . their having a central-nervous-system realization is not essential to them'. He further says that 'mental states are identical to contingently physical states [*sic*]'.[40] In these last two remarks Boyd *seems* to be saying that mental states *are identical* with the neural states that 'realize' them, although the identity is a contingent one (which would be a view like Armstrong's). But in the previous passage Boyd seems to be *rejecting* any identity between mental states and their 'realizations'.

Computational states. Let us leave the morass of 'realization' and take a brief look at the functionalist contention that all mental states are 'computational' or 'information-processing' states. Boyd says he wants to defend the view that 'mental events, states and processes are computational'.[41] Since he holds that all mental states are computational states and since, like other functionalist writers, he calls physical pain a 'mental state', Boyd will want to claim that pain is a computational state. Sure enough, that is exactly what he does claim. He says: 'mental states like pain are computational and are subject to the

[38]Ibid., p. 102.
[39]Boyd, 'Materialism without Reductionism', p. 102.
[40]Ibid., p. 105, n. 15.
[41]Ibid., p. 96.

same potential mistakes regarding their essential features'.[42] Here again it is difficult to understand what the view is. If a man is worried about whether he can pay his bills, the worry might be the result of a computation. The worry is not a computation, but a computation enters into it. In contrast, if a person is in dreadful pain because he was struck on the head by a brick, what would a *computation* have to do with it? Does he cry out with pain because of a computation? Does he compute the severity of the pain? Might he make an error of computation and so cry out with pain *by mistake* (as Boyd's remarks suggest)?

The weird idea that all mental states are computational states, springs from an enchantment with computing machines. The states of a computer are computational states; people are like computers; ergo, the mental states of people are computational states. This is sorry reasoning. Anything can be said to be *like* anything else, in some respect or other. People and computers are alike in that people sometimes compute, and so do computers. But the two are also radically unlike. As was observed in section 1, machines cannot be said to be literally conscious or unconscious, and therefore states of consciousness cannot literally be ascribed to them. This observation alone is enough to dismiss the computer from the philosophy of mind.

4 Conclusion

It is remarkable that philosophers seeking an understanding of the mental concepts, have lost sight of *the bearer* of mental predicates. Descartes held that an invisible, intangible, immaterial mind is that which thinks, wills, suffers. Present-day philosophy has justifiably turned away from the Cartesian view, but has proposed instead something equally absurd, namely, that the human brain, or even the computational states of machines, are the bearers of mental predicates. It is as if philosophers *could not believe* that the living corporeal human being is the subject of those predicates.

Human life contains many elements or stages: birth, childhood, family life, schooling, sexual awakening, love, friendship,

[42]Ibid., p. 98.

marriage, work, poverty, parenthood, ageing, illness, death. These destinies and vicissitudes are undergone and suffered by *people*, by you and me: *not* by immaterial minds or brains or machines. The human being who encounters those conditions is *the subject*—who can be said to be ignorant, timid, conceited, ambitious, affectionate, greedy, generous, honest, despairing, devoted, ungrateful, resentful, resolute.

Our application of those terms to our fellows is not guided by the examination of neural processes, but by our knowledge or belief about the situations and conditions in which those people are placed. In relation to such contexts we may interpret a person's countenance and behaviour as betraying jealousy of a rival; we see a glance as timid, a smile as conceited, a look as resentful, a shrug as despairing. We may know enough about the situations in life of our acquantances and neighbours, to understand that they have *an occasion* for anger, or anxiety, or gratitude, or despair. We are often unsure or mistaken. But our attributions of these attitudes, emotions, feelings, to people, only make sense in relation to their interests, concerns, engagements, family ties, work, health, rivalries—and make no sense at all as attributions to disembodied minds, or to brains or machines.

It is astonishing that philosophers who, in a sense, know perfectly well that mental terms apply primarily to living persons, like themselves, should propound theories according to which joy, fear, surprise, regret, dismay, are states of immaterial minds, or of brains, or of automata. These philosophers in their everyday life employ the mental terms as do the rest of mankind—applying them first of all to living human beings, and applying some of them by analogy to other animals. But when philosophers step from ordinary life into their studies, they become bewitched. They no longer understand what they have always known. They no longer see what is and has always been in plain view. We may say with Wittgenstein: 'God grant the philosopher insight into what lies in front of everyone's eyes.'[43]

[43]Wittgenstein, *Culture and Value*, edited by G. H. von Wright in collaboration with Heikki Nyman, translated by Peter Winch, Blackwell, 1980, p. 63.

Consciousness and Causality

D. M. ARMSTRONG

Acknowledgements

The following persons have read a draft of my reply, and given me much valued advice, a good deal of which I have taken or tried to take: Michael Devitt, Frank Jackson, David Lewis, Bill Lycan, Jack Smart and Kim Sterelny. I am also grateful to Cam Perry for advice about psychological literature.

1 Introductory

We all have a certain picture of the relation of a person's body to that person's mind, although it may not be the only picture which we have. According to this picture the body is a thing, a material object, a particularly remarkable and interesting material object. This body may be in one or another of a huge variety of physical states. A huge variety of physical events and processes go on in it. In some way set over against this body is the mind. The mind is also a thing, or at least it is an arena of some sort. (Hume pictured it as a theatre, although he went on immediately to say that the picture should not be taken seriously.)[1] This thing, or arena, may be in one or another of a huge variety of mental states, and a huge variety of mental events and processes go on in it. The body and the mind act causally upon each other. A blow upon the hand, a physical event, may cause pain, which is a mental state. A train of thoughts, a mental process, may cause sounds to issue from lips, a physical process. In the technical language of philosophers, the picture is *interactionist*. It permits, indeed requires, two-way causal action between body and mind.

I will speak of this picture as the traditional picture.

In my view, we all have this picture because it is, by and large, a correct one. Mankind has had abundant opportunity to observe, theorize and meditate on this topic. Through the ages they have found this picture forcing itself upon them almost irresistibly. I think the moral is that the picture embodies a true theory. At any rate, it is the account which I will defend in what follows.

As I have briefly set it forth, the picture, or theory as I shall henceforward call it, is somewhat vague about the exact nature of the mind. Concerning this nature, the theory can be developed in a number of ways. It might be thought that the mind is something completely different from the body: that it is non-physical, immaterial, spiritual perhaps. A view of this sort was upheld by Descartes, who argued that the mind was a

[1] David Hume, *A Treatise of Human Nature*, edited by L. A. Selby-Bigge, Oxford University Press, 1973, book I, part IV, sec. VI, p. 253.

spiritual substance. A different version of immaterialism was maintained by Hume. He thought that the mind was a collection or bundle of non-physical, non-spatial, phenomena ('impressions' and 'ideas'). But one is not compelled to accept one of these *dualist* theories of the mind. One can accept the general picture of the relation of body to mind which I began by outlining, yet hold that the mind is *part* of the body. The Greek atomists, for instance, thought of the mind as a collection of especially smooth, round, fiery atoms which coursed around the collection of more ordinary atoms which constituted the rest of the body. Contemporary *materialists*, or *physicalists*, however, with the benefit of many centuries of accumulated scientific knowledge and plausible theory, identify the mind with the brain, or, in more systems-oriented language, with the central nervous system.

I will defend the traditional theory in its materialist form.

There are those psychologists and philosophers who wish to reject the traditional theory, both in its dualist and materialist (and any intermediate) form. Among psychologists there have been the behaviourists J. B. Watson and, more recently, B. F. Skinner. Among philosophers there have been Ludwig Wittgenstein and Gilbert Ryle. The author of the first essay of this volume, Norman Malcolm, is a distinguished disciple of Wittgenstein. So I will have to defend the traditional theory (in its materialist form) against Malcolm.

However, I think that these thinkers do have a legitimate complaint against the traditional theory, or at any rate the theory as it has been developed by a number of its philosophical defenders. There seems to be a closer link between the mind and some of the outward physical behaviour of the body than the traditional theory allows for. In the traditional view, the mind is locked away inside, either literally inside the body, as materialists assert, or metaphorically, as dualists hold. But does not (some) outward physical behaviour come closer to the essence or definition of the mental than this? Consider this stone lying on the ground. Does it perhaps have some sort of mind and, if so, could it be that it is in great pain? Wittgenstein writes:

> Look at a stone and imagine it having sensations.—One says to oneself: How could one so much as get the idea of ascribing a

> *sensation* to a *thing*? One might as well ascribe it to a number!—And now look at a wriggling fly and at once these difficulties vanish and pain seems able to get a foothold here, where before everything was, so to speak, too smooth for it.[2]

The difference between the case of the fly and the case of the stone appears to be that the fly's behaviour at least resembles pain-behaviour, while the stone's behaviour does not. But this in turn suggests that there is a connection, closer than mere causal connection, between pain and pain-behaviour. The pain-behaviour seems to be of the essence of pain. Yet this conclusion is barred by the traditional view, at any rate as it has often been developed.

Actually, every muscle in the body can be totally paralyzed, so that one is unable to exhibit any behaviour at all, and yet one can be in great pain. (The drug curare can be used to bring about such a paralysis, a paralysis which need not affect sensation and consciousness.) So it might be objected to Wittgenstein that the stone could be in pain but lack any means of expressing that pain in its outward behaviour. However, even in this case it can still be said that there is a natural outward expression of pain which the stone would give expression to *if* it was able to. Hence it still seems plausible to say that there is some essential or definitional link, even if a subtle one, between pain and the natural behavioural expression of pain.

It is customary to discuss this point in connection with mental items such as pains. But it is even clearer in the case of conative mental states, such as purposes. If it is somebody's objective to open the door, then, by and large and in the absence of obstruction, door-opening behaviour will follow. But now suppose, as the traditional theory supposes, that the purpose is something in the arena of the mind, something which, in suitable circumstances, *causes* this behaviour. Add to this premise the plausible assumption that what causes what is a contingent matter, and that 'to consider the matter *a priori*, any thing may cause any thing' (Hume).[3] How then can it be

[2]Ludwig Wittgenstein, *Philosophical Investigations*, edited by G. E. M. Anscombe and R. Rhees, translated by G. E. M. Anscombe, Blackwell, 1953, §284. (Hereafter referred to as *PI*.)

[3]Hume, *Treatise*, book I, part IV, sec. V, p. 247.

of the logical essence of this cause that it tends to produce door-opening behaviour? As a result, there seems to be something unsatisfactory in treating the purpose as a cause in an inner realm. The purpose must be more closely tied to its behavioural expression than this.

I think that those, like Wittgenstein, who oppose the traditional theory have an important point here. But I will try to show that there is a version of the traditional theory which can accommodate their point without abandoning the traditional view. As it happens, this version is hospitable to, although it does not logically entail, materialism.

Summing up: (1) I will uphold the traditional view of the relation of body to mind; (2) I will uphold it in a materialist version; (3) nevertheless, I will concede that there is something to be learnt from the criticism of the traditional view to be found in Wittgenstein and Ryle. In expanding these views, I will at various points take issue with what Malcolm has said in the first part of this book.

2 In Defence of Inner Sense

An epistemology for the traditional theory

Associated with the traditional theory is an account of how persons come to be *aware* of bodies and of minds, and, in particular, how persons come to be aware of their own minds.

It is clear that we gain such knowledge as we have of the current state of the physical world by means of sense-perception. By the senses we gain information about the state of our body, the state of its material environment, and the relations in which the body stands to this environment. The information is mixed in with a certain amount of misinformation, much of which, however, we learn to correct for.

So much is uncontroversial. However, the traditional theory goes on to make a further assertion which has been questioned. It asserts that for the case of a person's own mind, though not for any other mind, a perception-like process goes on which yields the person information about the current state of his mind. A classical exposition of the theory is to be found at the beginning of Book II of John Locke's *Essay concerning Human Understanding*.

After putting forward his central doctrine that all our ideas come from experience, Locke says that there are two sources of experience, two 'Fountains of Knowledge':

> First, *Our Senses*, conversant about particular Objects, do convey into the Mind, several distinct Perceptions of things, according to those various ways, wherein those Objects do affect them: And thus we come to those *Ideas*, we have of *Yellow, White, Heat, Cold, Soft, Hard, Bitter, Sweet*, and all those which we call sensible qualities . . . This great source, . . . I call SENSATION.
>
> Secondly, The Other Fountain, from which Experience furnisheth the Understanding with *Ideas*, is the *Perception of the Operations of our own Minds* within us, as it is employed about the *Ideas* it has got; which Operations, when the Soul comes to reflect on, and consider, do furnish the Understanding with another set of *Ideas* which could not be had from things without: and such are, *Perception, Thinking, Doubting, Believing, Reasoning, Knowing, Willing*, and all the different actings of our own Minds; . . . This source of *Ideas* every Man has wholly in himself: And though it be not Sense, as having nothing to do with external Objects; yet it is very like it, and might properly enough be called internal Sense. But as I call the other *Sensation*, so I call this REFLECTION, the *Ideas* it offers being such only, as the Mind gets by reflecting on its own Operations within it self.[1]

Locke's view is reproduced by Kant. What Locke calls 'sensation', Kant calls 'outer sense'. Echoing Locke's own phrase, Locke's 'reflection' becomes 'inner sense':

> . . . by means of which the mind intuits itself or its inner state.[2]

[1]John Locke, *An Essay concerning Human Understanding*, edited by P. Nidditch, Oxford University Press, 1975, book II, ch. I, secs. 3–4.

[2]Immanuel Kant, *Critique of Pure Reason*, translated by N. K. Smith, Macmillan, 1950, A23/B37.

Those contemporary philosophers and psychologists who accept the Locke–Kant view speak of 'introspection', 'introspective awareness', or, in one important sense of a very vexed term, 'consciousness'. Malcolm in his essay (p. 25) points out that the doctrine was held by the philosopher Franz Brentano. But, indeed, it is philosophical and psychological orthodoxy. It is true that during the period when behaviourist modes of thinking were dominant, from the thirties to the fifties of this century, the doctrine fell into some disrepute. But now it is back in favour.

Reflection, inner sense, introspection, introspective awareness, consciousness (in one sense of the word), whatever we call it, can seem a very mysterious phenomenon. But I believe that there exists a demystifying, naturalizing, *model* for it in the field of perception. I refer to *bodily perception*, or, as some psychologists call it, *proprioception*. It will be appropriate to spend a little time discussing this interesting form of perception.

It is, or was, customary to speak of the 'five senses': seeing, hearing, tasting, smelling and touching. But our sensory capacity goes beyond these. Consider our awareness of our own bodily temperature. We have a certain amount of direct awareness of it. We are often aware that our temperature is up or down, or that parts of our body are abnormally hot or abnormally cold. Yet we may not need to use any of the five senses, even the sense of touch. Again, we have awareness of the general position of our limbs, and of the motion of our limbs ('kinaesthetic perception'). We are aware how our body as a whole is placed in relation to the earth: whether it is off-balance, upside down or normally placed. The 'five senses' co-operate with bodily perception to give us such awareness, but there is an independent proprioceptive contribution. Bodily perception has a spectacular type of illusion associated with it: the phantom limb.

So, over and above the five senses, there is awareness of bodily temperature, awareness of position and movement of limbs, and awareness of the relation of the body to the earth. We bring these forms of awareness together as bodily perception, or proprioception. The matter is more controversial among philosophers, at least, but I would argue that sensations of pain, itch, tickle, etc. are also bodily perceptions.

They are perceptions of idiosyncratic disturbances at various places in the body. These perceptions, like any other perceptions, can be illusory in greater or less degree.[3]

It might be argued that these forms of bodily awareness are not really senses in the way that vision, hearing, taste, smell and touch are senses. For in the latter cases there are sense-organs connected with each sense, organs which must be stimulated for perception to occur, and which we can manipulate in various ways to produce perceptions. It may be replied to this, however, that physiological research has located temperature receptors for heat and cold, receptors in the joints for posture and kinaesthetic perception, the Eustachian canals near the ears for states of balance, and so on.

It is true that we cannot *manipulate* these receptors in the way that we can manipulate our eyes, ears, tongue, nose, hand and other touch-sensitive areas. But this does not seem to be a big enough difference to justify restricting the term 'sense-perception' to the deliverances of the traditional five. In any case, manipulation of an organ demands awareness of the state of the organ, to serve as informational feedback. If this awareness in turn always involved the use of a manipulable organ, then a vicious regress would loom. Bodily perception is the sense which, *par excellence*, informs us of the state of manipulable organs. So we can actually expect that it itself would not operate by means of manipulable sense-organs.

It is true also that it is in many cases difficult to pick out the sensations involved in bodily perception. I know, without using any of the five senses, that my legs are now crossed. But do I have a sensation as of my legs being crossed? And, if not, am I *perceiving* that my legs are crossed? However, I think that in fact sensations are always involved. They are not conspicuous because bodily perception is usually background perception, and so not the object of much attention. It takes abrupt change, unusual states, or striking illusions to bring out clearly the sensational/perceptual element.

However, my interest here in bodily perception is not so much for its own sake, but rather as a model for introspective awareness. From this point of view, the interesting thing about

[3]See D. M. Armstrong, *A Materialist Theory of the Mind*, Routledge & Kegan Paul, 1968, ch. 14.

bodily perception is that, with respect to it, each person is confined to his or her own body. A can use A's five senses to gain knowledge of the current location, state, etc. of B's body. But A's bodily perceptions will not, save by some eccentric inferential route, assist in this task. A can also use his five senses to gain knowledge of the location, state, etc. of A's body. In addition, however, A will have his bodily perceptions. As a result, A will have a certain (strictly limited) epistemological privilege with respect to A's body. So for B with respect to B's body, and so for everybody else.

A person has a route of epistemological access to his own body which others lack, although they have the same privilege concerning their own bodies. So it should not be too mystifying that a person has a route of epistemological access to his own mind which other persons lack.

It might be objected that there are no sense-organs, nor even receptors, connected with introspective awareness, so that, at the very least, the Locke–Kant metaphor of an inner sense limps badly. Certainly there seems to be no sense-organ, no internal apparatus, to be *manipulated* in order to yield introspective awareness, in the way that the eyes are turned to yield visual information. Perhaps there are not even receptors, in the sense that there are proprioceptive receptors. But the materialist, at least, will assume that there are mechanisms, of a more or less complex sort and involving a variety of causal links, by means of which the mind becomes aware of some of its own current states and processes. That seems enough to sustain the value of the analogy.

This does *not* mean that only I can know what is going on in my mind. Others can know a good deal. In particular, they can come to know a great deal about my purposes, especially my short-term and unsophisticated purposes. They can come to know what my emotions are, sometimes identifying them better than I can. They can come to know the content of my perceptions, at any rate if that content is characterized in a rough-and-ready way. They can come to know some of the things which I believe.

This knowledge which others can have of my mind is linked with the one point where I would criticize Locke in the passage recently quoted. Locke says that the 'ideas' of the mental operations 'could not be had from things without'. In the next chapter it will become clear why I disagree with Locke here. But even if, as I believe, Locke has overstated his case in this respect, it still seems evident that each of us has *an* access to his or her own mind that all others lack. Think of dreams, of the having of mental images, of idle thoughts. In such cases we often know what is passing through our minds, but it is entirely optional whether we inform others. And unless we do inform them, they will normally be totally ignorant of these goings-on. (I apologise for the obviousness of this, but, as Wittgenstein noted, an important part of philosophy is the assembling of reminders.) The Locke–Kant–Brentano theory explains the epistemological asymmetry admirably.

Although, in bodily perception, each of us is confined to an awareness of states of, and happenings in, our own body, it does not seem difficult to envisage this restriction being removed. We can conceive being directly hooked-up, say by a transmission of waves in some medium, to the body of another. In such a case we might become aware e.g. of the movement of another's limbs, in much the same sort of way that we become aware of the motion of our own limbs. In the same way, it seems an intelligible hypothesis (a logical possibility) that we should enjoy the same sort of awareness of what is going on in the mind of another as the awareness we have of what is going on in our own mind. A might be 'introspectively' aware of B's pain, although A does not observe B's behaviour.

If the case of pain is taken, there are complications because of the complexity of the phenomenon of pain. It is a species of bodily sensation which characteristically evokes a very direct desire that the sensation should cease. If the fantasy of direct awareness of the pain of another is admitted as logical possibility, as I have just suggested that it should be, then it could be developed in (at least) two different ways. First, the direct awareness of pain in another's limb might be accompanied by the same reaction which one has to one's own pain. The limb of the other would be treated in this respect as if it were an extension of one's own body.

But, second, the awareness of the pain might generate, at best, the relatively abstract sympathy which one has for the sufferings of others. There are, indeed, cases where dissociation between characteristic sensation and characteristic reaction occurs even in the first-person case. Malcolm (p. 12) notes that there are cases where, after surgical or other treatment, patients report sensations of pain but are largely indifferent to the having of the sensations. Malcolm finds such cases hard to comprehend, but I find his incomprehension puzzling. For surely physical pain involves at least two elements: (1) bodily sensation of a certain sort or sorts; (2) an unfavourable reaction to the having of these sensations. (Contemporary psychology distinguishes in a fairly routine way between the bodily sensation of pain, on the one hand, and the distress or suffering it causes, on the other.)[4] But if pain involves these two elements, why should not the two be dissociated in some special circumstances? Ordinary language, of course, is not well equipped to deal with such unusual cases.

It may be thought that 'introspective' awareness of the pain of another could not be compared to our awareness of our own pain, because of the epistemological asymmetry involved. For suppose that I claim such direct awareness, telepathic awareness, perhaps, of your pain. Will not my claims have to be checked against your admissions and/or behaviour? And even if I turn out to be remarkably successful in my claims, does this not mean that your access to your pain is of a quite special authority, and so is access of a quite special nature? The person himself is the ultimate authority. So the successful telepath only has borrowed authority.

At this point the dispute threatens to become entangled with the question whether inner sense is an infallible sense. Here I will say only that I shall be discussing, and rejecting, this doctrine of infallibility below (see pp. 135ff). Introspective awareness is in some sense direct awareness, but such direct awareness does not have to be infallible awareness. It can be, if I may so express myself, a false awareness. Once we recognize this point, we see that the special authority that a person has about his own current mental states, by comparison with a hypothetical (logically possible) 'direct observer' of these same states, is of no great moment for the following reason.

[4]For instance, E. R. Hilgard, *Divided Consciousness*, John Wiley and Sons, 1977, p. 193, pp. 246–7.

New ways of gaining knowledge of a certain range of phenomena must in the first instance be tested by checking them against older ways of gaining knowledge of the same phenomena which have already proved themselves reliable. Introspection is a reasonably reliable way of gaining direct knowledge of some features of some of our own mental states. Successful claims by other persons to have such direct knowledge, which we are imagining to occur, would therefore have less *initial* authority than introspective awareness. However, there is no reason why such claims by others should not acquire equal or even greater authority *after they had proved themselves*.

Having the same pain

In the imaginary cases just considered, one person has a pain, and is introspectively aware of it in the usual way. The imaginary feature is that another person is aware of that very pain in the same unmeditated way that the person who has the pain is aware of it. It may further illuminate the position defended in this essay if we consider a further imaginary case where two persons have numerically the same pain, are both introspectively aware of it and are both distressed by it.

It is sometimes suggested that Siamese twins could have the very same pain. Let there be an area which is part of both bodies, let it be suitably linked to the central nervous system of each, and then let a pin be stuck into this area of flesh. Each will feel a pain, and if asked 'where it hurts' will nominate numerically the same place. But are the pains which they feel numerically the same? Given that pains are bodily sensations, and that sensations are affections of minds, then twin A's being in pain will be an affection of A's mind and twin B's being in pain an affection of B's mind. As a materialist, I would take the two pain-states to be affections of the two central nervous systems. If so, there are in this case *two* pains, not one, although, of course, they may be exactly the same sort of pain.

What is numerically one is the place where the fleshly disturbance feels to each of them to be (and in the given case actually is). The potentially misleading idiom whereby this place is referred to as 'the place of the pain' may be responsible for the idea that there is just one pain. But there are certainly two pains. Aspirin administered to twin A might take away his pain, but leave twin B unaffected. The twins are in no different situation from two people who see, or seem to see, a blue ball at exactly the same spot: two perceptions, but at most one ball.

What we want instead is a case of Siamese minds. For the materialist about the mind, this will come down to a case of Siamese brains, and it is convenient, although not essential, to develop the case in materialist terms. Suppose, then, that two persons have an overlapping portion of brain and that certain processes, states, etc, in the overlapping portion are functionally part of *both* minds. Suppose that all havings of sensations of pain in both minds are processes in this Siamese portion. Whether they know it or not, the twins really will have numerically the same pains. Suppose, however, that the portions of their minds which react to pain sensations, including their introspective apparatus, have no such overlap. There will be two cases of distress, and two awarenesses of the one pain.

Consciousness

It will now be useful to look at the notion of consciousness. It appears that we can distinguish a number of different concepts of consciousness.

A person can (perhaps) be totally unconscious. Then they can 'regain consciousness'. What is it to regain consciousness? Malcolm introduces a useful distinction here. He speaks of consciousness in this sense as *intransitive* consciousness. His reason is that we can understand what it is for somebody to regain consciousness without having any idea of something that the consciousness is a consciousness *of* (transitive consciousness).

However, Malcolm goes on to make a suggestion which I think is very dubious. His suggestion is that it is possible to have intransitive consciousness without having transitive consciousness:

> A person who has been knocked unconscious could show signs of returning consciousness (moving his arms, opening his eyes, sitt-

> ing up, muttering) without its being apparent that he is conscious *of* any sights, sounds, sensations or anything at all. (p. 30)

I do not know how to disprove this contention of Malcolm's, but I see no reason to follow him here. To be sure, to say that a person has regained consciousness in no way tells what he is conscious of. But I take it that it is the special usefulness of an expression like 'He has regained consciousness' that its meaning is indeterminate with respect to the content of consciousness. The behavioural signs are that he has regained consciousness. But it is hard to say just what he is conscious of. So one wants an intransitive *idiom*. But this is compatible with him being conscious *of* something. And I think that in fact all consciousness is transitive, that is, it is consciousness of something. (This is not to say it is always consciousness of something that *exists*, simply that the consciousness will always have a content.)

A parallel here is to be found in our perceptual language. Suppose it is true to say of somebody that he is seeing a horse. Normally at least, this is an intransitive idiom. It is true that it entails that the horse (a certain portion of the horse's surface) is acting on his eyes and causing him to have certain perceptions, perceptions of the horse. It is true also that these perceptions will have a detailed content, in particular, the object seen will be sorted and classified in a certain way. But what is this content? Just by considering the statement that he is seeing a horse, we cannot say. It may be, for instance, that he quite fails to classify the object as a horse. He may perceive it as nothing more than a thing of a certain shape and colour, and even these shapes and colours may be misperceived.

Nevertheless, the perception will have some more or less determinate content. Godfrey Vesey has said that all seeing is seeing *as*,[5] which I understand as the correct claim that all seeing involves sorting and classifying. This can be extended to all perception (and to 'inner perception' also). Sorting and classifying, at the sub-verbal level, is always present. There are intransitive perceptual idioms, but there is no intransitive perception.

An intransitive idiom like 'He saw a horse' is often useful to us, useful just because it is indeterminate. We often know that

[5] G. N. A. Vesey, 'Seeing and Seeing As', *Proceedings of the Aristotelian Society* 56, 1956.

somebody has seen something without knowing the detailed nature of his visual perceptions. Again, the intransitive idiom may be useful to the perceiver himself. He can say that he saw a horse without undertaking the arduous task of specifying his visual perceptions in any more detail.

In the same general way, I suggest, to say that somebody has recovered consciousness entails that he is transitively conscious in some way, without in any way specifying the nature of that consciousness.

There is a complication here, though. What are we to say about a person 'in a sleep-walker's trance', as Malcolm puts it (p. 31)? Are we to say that such a person is conscious, or not? And if he is conscious, can he really be said to be *transitively* conscious? I may add that I am not clear what Malcolm thinks about such a case, although he does speak of 'degrees of consciousness' in this connection.

I believe that we can cast light on cases like that of the sleep-walker, and in so doing further illuminate the doctrine of inner sense, if we distinguish three types of consciousness, or, if you like, three senses of the word 'consciousness'.[6]

The totally unconscious person has a mind. Things can be truly said about that mind. In particular, beliefs and purposes can be attributed, although beliefs cannot be manifested in any way, nor can any progress be made towards carrying out the purposes. (I find the image of a switched-off computer useful here. Its memory bank may be loaded, and it might even be in the middle of carrying out some routine.) In the totally unconscious person there are no mental processes occurring, just as there are no computing activities occurring in the switched-off computer. It may be that *total* lack of mental activity is a limit, never actually reached while the mind exists. But that, I take it, would be a matter for determination by psychologists and/or neurophysiologists.

Now suppose that this unconscious person begins to dream. I assume, *contra* the opinion of Malcolm,[7] and perhaps Wittgenstein,[8] that dreaming is something which goes on at definite

[6]See D. M. Armstrong, 'What is Consciousness?' in D. M. Armstrong, *The Nature of Mind*, Harvester Press, 1981.

[7]Normal Malcolm, *Dreaming*, Routledge & Kegan Paul, 1959.

[8]Wittgenstein, *PI*, p. 184.

times during sleep. It is a mental *activity*. (The correlations established between dreaming and REM—rapid eye movement —sleep seems to put this beyond doubt.) As a result, we can say that the person is not totally unconscious. This gives us a first sense of the word 'consciousness'. *In this sense* a person is conscious at *t* if and only if there is mental activity in his mind at *t*. We may call this *minimal* consciousness.

Persons who are dreaming are in a sense conscious. But if they are sleeping soundly, they are not in any noticeable degree conscious of their environment. They are 'dead to the world'. In particular, they do not appear to *perceive* what is going on in their environment or, for that matter, in their own body.

I suggest that this gives us a second sense for the word 'consciousness'. In order to be conscious in this second sense, it is necessary and sufficient that the persons be perceiving the physical world. (Perceiving here is not meant to be restricted to *veridical* perceiving of the world.) The dreamer in sound sleep lacks consciousness in this sense, or so we may assume. Let us say that what he lacks is *perceptual* consciousness. It will be seen that perceptual consciousness entails minimal consciousness, because perception is a mental activity.

We can now consider 'the sleep-walker's trance'. One of the known facts about sleep-walkers is that they go through quite complicated physical routines. Such routines are impossible of accomplishment unless the persons carrying them out can get continuous feedback about the state of their body and its relation to its environment. (Consider a sleep-walker successfully walking down the stairs.) It seems, therefore, that sleep-walkers must be *perceiving* while they are walking. The 'must' here is not the 'must' of logical necessity, but rather that of good scientific inference. It seems that in order to *explain* how sleep-walkers are able e.g. to walk down the stairs we must assume that they are perceiving. If so, then, by our definitions, they have perceptual, and so minimal, consciousness.

At the same time, however, sleep-walkers appear to be quite unaware of what they are doing. Not only do they have no later memory of their actions, but they appear to have no consciousness of their actions at the time that they are performing them. Yet they have perceptual consciousness. If, then, they 'lack consciousness', it must be consciousness in some further sense which we have not yet defined.

I suggest that, in all likelihood, what is going on in sleep-walkers is that their inner sense is not functioning. They lack the power of Reflection. They are not introspectively aware. Although they are perceiving, and although, clearly, they have purposes of sorts, if only the purpose to walk down the stairs, they appear not to be aware that they are perceiving and appear not to be aware of what it is that they are aiming at. They appear to lack *introspective consciousness*.

It may well be that Malcolm's man who is beginning to regain consciousness but is not yet in a normal state, also lacks introspective consciousness. Of course, we cannot be more than tentative here, any more than we can in the sleep-walking case, or even in the case of ordinary dreaming, because empirical investigation of such cases might lead us to say something else. It is hard to find secure paradigms in this tricky field. But for the purpose merely of introducing a certain concept, we can take the cases at face value.

In ordinary waking life, however, we not only have perceptual consciousness, but we are also introspectively aware of our perceivings, along with much else which passes in our minds. We have introspective consciousness. Introspective consciousness is, *ipso facto*, minimal consciousness. But it does not actually seem to *entail* perceptual consciousness. A person who was having no perceptions might still be introspectively aware of other mental goings on.

A further distinction is now required. In vision we can distinguish between careful scrutiny, on the one hand, and idle, passive, reflex, merely recipient, gazing, on the other. The same distinction, it appears, can be made in the case of inner sense. Normally, introspective consciousness is of a pretty relaxed sort. The inner mental eye takes in the mental scene, but without making any big deal of it. (Compare bodily perception in normal circumstances.) It is, however, possible to undertake introspective scrutiny, to bend one's energies to try to discover the exact nature of what is going on in one's mind. As one would expect, some people seem to be better at it than others.

One thing which it seems possible to become introspectively conscious of is our own introspective consciousness. We can certainly say, apparently truly, 'I am now attempting to observe my own current mental goings on'. How can we be aware of this? Only, it seems, by observing our observing of our own mind.

Descartes's myth

Ryle called the first chapter of *The Concept of Mind* 'Descartes' myth'. For him the myth was the traditional two-thing theory of the relation of body and mind, in particular the dualist form of this theory espoused by Descartes, and, before him, Plato. (But he thought that the materialist identification of mind and brain was little better.) I maintain that Descartes's view is not mythical at all, although I hold that his dualism is mistaken. Nevertheless, I do think that Descartes left us with a myth, a myth which has done untold damage in the philosophy of mind and in the study of mind generally, a myth from which our culture is only now shaking itself free. It is a myth about inner sense, and, explicably, it is an exalting of inner sense. But by an irony of a not uncommon sort, its result has been to discredit the doctrine of inner sense and, as a further result, to discredit the traditional theory of the relation of mind and body.

Descartes held that the essence of mind is consciousness. By consciousness he meant in the first place mental activity: what I have called minimal consciousness. It was this doctrine which earned the scorn of Locke:

> . . . methinks, every Drowsy Nod shakes their Doctrine, who teach, That the Soul is always thinking [is mentally active].[9]

I would say that Locke is at least so far right: it is not a necessary truth that the mind is always mentally active. In sleep, in particular, it could at times be like a stopped engine.

But Descartes did not merely think that mentality involves minimal consciousness. He took it that it always involves introspective consciousness. Everything in a mind at a time is available to consciousness, introspectively available, at that

[9]Locke, *Essay*, book II, ch. I, sec. 13.

time. We may speak of this view as the doctrine of 'self-intimation'. The doctrine itself became philosophical orthodoxy. Although Locke, for instance, thought that the mind could be mentally inactive, he also thought that when the mind *is* mentally active, then the activities of the mind automatically lie open to reflection. This is why Locke says, as quoted by Malcolm (p. 23), that it is:

> impossible for any one to perceive, without perceiving, that he does perceive. When we see, hear, smell, taste, feel, meditate, or will anything, we know that we do so.[10]

Here Locke, like most Western philosophers until quite recently, accepts the Cartesian doctrine of self-intimation. (Leibniz was a shining exception.) Among other things, this retarded the recognition of the reality of unconscious mental processes.

The doctrine of self-intimation ensures that whatever is in the mind at *t*, we are aware at *t* that it is in the mind. Nothing escapes inner sense. But it is logically compatible with holding that inner sense may be erroneous. If A is in mental state M at *t*, then A is aware of being in M. But A might be 'aware' of being in state N at *t*, although in fact A was not in state N at all. Self-intimation might hold, yet inner sense be an overreacher.

But this too Descartes denies. He assumes that introspective awareness is infallible. This comes out in his *Second Meditation*, when he considered how to answer the sceptical arguments proposed in the *First Meditation*, and in particular the argument from the possibility of a deceitful demon. He says that the demon can bring it about that he, Descartes, is not, as he thinks he is, sitting in front of a fire and feeling its heat. But even the deceitful demon cannot bring it about that he thinks he has visual perceptions as of fire when in fact he has no such perceptions, or that he thinks he has sensations of heat when he has no such sensations. His awareness of his own perceptions is logically indubitable, that is, it is logically impossible that he should be mistaken. And, it appears, Descartes wishes to claim of all consciousness which restricts itself precisely to the current mental state that it is logically indubitable. The deliverances of inner sense are *self-evident*. Like the doctrine of self-intima-

[10] Ibid., book II, ch. XXVII, sec. 9.

tion, the doctrine of self-evidence became philosophical orthodoxy.

Self-intimation and self-evidence are brought together by David Hume in a remarkable passage:

> Since all actions and sensations of the mind are known to us by consciousness they must necessarily appear in every particular what they are, and be what they appear.[11]

Nor is Hume's assent to the Cartesian twin principles notional only. He makes plentiful use of the principles in the course of his philosophical reasonings.

So for the Cartesian tradition the current contents of the mind all intimate themselves to an infallible faculty. What glory for inner sense! But, of course, to many, especially to those with a naturalist bent, it all seems too good to be true. The proper reaction, I believe, and will argue in detail shortly, is to scale down inner sense to a faculty like any other human faculty of knowledge, a faculty which fails to discern much, and which can fall into error. But another, not completely unjustified, reaction is to reject the notion of introspective awareness altogether. Exaggerated claims give rise to exaggerated counter-claims. The doctrines of self-intimation and self-evidence have led, in some philosophers, to the discrediting of the doctrine of inner sense.

Over-valuing of the power of introspection can lead to problems concerning our knowledge of other minds. In introspection, it appears, we are directly aware of some of the current contents of our own minds. We appear to have no such direct awareness of the contents of the minds of others. Could we have such a direct awareness? I have already argued (pp.113–15) that this is logically possible. But one thing seems perfectly clear. Such a hypothetical awareness would be subject to error, it would be correctable. If, however, our awareness of our own mental contents is an infallible one, then our assurance of the existence of other minds, even in the best conceivable circumstances, must fall short of our assurance of the existence of our own. A certain scepticism about the existence of other minds is then hard to avoid.

[11] Hume, *Treatise*, book I, part IV, sec. II, p. 190.

I wish to defuse this problem by denying that introspection is logically infallible. But if we assume that introspection, if it exists at all, *is* infallible, then we may seek to avoid the problem in another way. We may conclude that there is no such thing as introspection, and, as a result, reject the whole view of mind as an inner realm.

I do not know why, in the past, the Cartesian version of the doctrine of introspection has had such a strong hold. I suspect it to be connected with the Cartesian quest for absolute certainty, for unimpugnable foundations for all knowledge, which has been such a feature of the philosophizing of the past few centuries. This desire for indubitable certainty is itself perhaps a response to the social and personal uncertainties which have so gripped European culture in the great age of what the Marxists rather crudely call 'bourgeous individualism'.

At any rate, if one *is* looking for unimpugnable foundations, there is a lot to be said for trying to find it in introspective awareness. If we could all have direct 'introspective' access to the minds of others, then the intersubjective checks on ordinary introspection which would then exist would no doubt disclose that we quite often fall into error about our own current mental states. But lacking this check as we do, it is possible to protect the hypothesis of introspective infallibility against any easy falsification!

Under the next two headings, however, I will argue that we have no good reason to accept the doctrine of self-intimation or of introspective infallibility.

Against self-intimation

A thing or phenomenon may not be seen, and yet be there to be seen in the field of vision. In these circumstances, all that is necessary for the thing or phenomenon to be seen is that it become the object of some visual attention. Alternatively, a thing or phenomenon may not be seen, either because it is not in the field of vision at all or because, although it is in the right place, it is not the sort of thing that can be seen by the perceiver. If introspective awareness is real, and can be thought of as the operation of an inner sense, then it should be possible to draw similar distinctions. There should be some current mental phenomena which we are not aware of, but of which we can make ourselves aware by suitably directing introspective attention. And there should be other current mental phenomena of which we are not aware, and of which we cannot make ourselves aware merely by the redirection of attention. I believe that plausible instances of both sorts of case can be found.

We first want cases where something mental is, as it were, in one's introspective field of view, but no introspective attention is given to it. I think that Malcolm provides such a case. He considers at some length a case where his attention is distracted from pain. His legs begin to ache during a long walk, but as a result of a lively conversation he ceases to be aware of the aching, only to become conscious of it again when the conversation stops. It seems to me that this is the sort of case required. The natural thing to say about this case is that the aching continued throughout the conversation, but that during that time Malcolm was unaware of the pain.

Malcolm does not want to say that during this time the ache stopped. However, he does not want to say that it continued either! He thinks that here the law of excluded middle fails. For:

> . . . we have no understanding of what it would be to investigate whether the aching continued or whether it stopped during the time that I was in the conversation. (p.15)

This is a puzzling remark. Malcolm appears to be trying to explain it shortly afterwards where he discusses what he calls 'indeterminacy'. As applied to the case in hand, his view seems to be this. It is imaginable that everybody, or practically everybody, when faced with such a case, would have spontaneously agreed that Malcolm was *not* in pain during the conversation. (After all, he was in no trouble.) This agreement would have fixed part of the boundary of our concept of an ache: the concept would have made no provision for aches of which we are unaware. Alternatively, it is imaginable that everybody, or practically everybody, when faced with such a case, would have spontaneously agreed that Malcolm *was* in pain during the conversation. (After all, if at any time he had stopped to consider the matter, he would have been aware that his legs were aching.) This agreement would again have fixed part of the boundary of our concept of an ache: but this time the concept would definitely provide for aches of which we are unaware.

As things are in fact, however, Malcolm thinks that there is no general agreement about how we should speak in a case like this. People are pulled two ways, although perhaps some ultimately go one way, some another. This has the result (indeed constitutes the fact) that the concept of ache is vague, or indeterminate, with respect to aches which we are not aware of as the result of distraction. For Malcolm it is a conflict case, like the case where we wonder whether chess minus the queen should be called 'chess' or not.

Thus Malcolm. It seems to me, on the contrary, that it is natural (although not inevitable) to conclude that Malcolm's limbs were aching all the time. Here is the argument.

One thing that Malcolm could have done at any time during the conversation was to stop and consider whether his limbs were aching or not. Presumably, if he had done this, he would have discovered that they were aching. But, if so, were they not aching all the time? It might be said that only a weaker conclusion should be drawn: that the aching exists whenever Malcolm stops to examine his bodily sensations. Perhaps at other times it goes away. But while its going away is a logical possibility, there is no reason to believe it occurs. Admittedly, it is harder to disprove this hypothesis than it is to disprove the hypothesis that a certain physical object ceases to exist when Malcolm ceases to observe it. For in the latter case Malcolm can ask another person to check the situation for him. The restricted scope of inner sense being what it is, this procedure would be unavailable for the ache. But if it is agreed that any random spot check would reveal to Malcolm that his legs were aching at the time of the check, then that is a good reason to think that they were aching during the whole conversation. For why should we suppose that making the check brings the ache into existence?

Malcolm resists this conclusion. The reason he gives is that we do not understand what it is to investigate whether the aching continued or whether it stopped during the time he was engrossed. He could back this up by pointing out that I have already conceded to him that nobody else can check the situation in the way that he can, and, by hypothesis, that he is not checking it because he is engrossed in conversation.

It seems, however, *contra* Malcolm, that investigations of a more indirect sort are possible.

First, there might be behavioural evidence. Suppose that the aching of Malcolm's limbs was somewhat selective, and that his right leg ached while his left leg did not. And supposing that, during his engrossing conversation, Malcolm, contrary to his usual gait, favoured his right leg. A natural interpretation would be that his right leg ached, although he was unaware of its aching.

On pages 9–10, some time before his discussion of the aching limb, Malcolm remarks that if a person is not aware of any pain at all, then there ought to be an absence of any of the pain-behaviour that is natural to human beings. This remark of Malcolm's seems quite wrong. If one takes seriously the distinction between pain and awareness of pain, we will expect pain which fails to give rise to awareness of pain nevertheless to have *other* effects, including behavioural effects. Contrariwise, if no such effects can be discovered, that will weaken the hypothesis that the ache goes on at the time Malcolm is not aware of it.

Second, there may be physiological evidence. An aching limb is presumably in a subtly different physiological condition from one which is not aching. Or there may be physiological signs of minor distress in the body generally.

Third, a neurophysiology more advanced than any which we now have might yield evidence that processes generally associated with aching limbs were going on in Malcolm's brain all the while that the conversation continued. Malcolm does consider this possibility, but argues against it on the ground that the neurophysiological data might still leave us with uncertainty and disagreement. So the data might. So might the behavioural data and more ordinary physiological data. This disagreement and uncertainty might be about the facts. But, and I suppose that this is the situation that Malcolm is thinking about, the relevant facts might be clear enough but there might be disagreement and uncertainty about how to speak about the facts. It might be like 'chess without the queen'. In such a case, it might have to be concluded that no phenomenon exists which can clearly be described as an ache without awareness of aching. But notice that, while this is a possible upshot, the neurophysiological and other data might also point rather unambiguously to aching without awareness of aching.

I do not claim to have proved my point against Malcolm. But I hope that I have shown that in his discussion of the aching legs Malcolm simply assumes that no very large wedge can be driven between aching and awareness of aching. However, if we assume instead that the conceptual distinction *can* be made, then it is easy enough to see what evidence would count in favour of, (and what evidence would count against), the actual existence of aching of which we are unaware. Of course, the evidence will not be logically decisive. But why should it be expected to be logically decisive? The quest for logically decisive evidence has been the bane of philosophy at least from the time of Descartes to the present day.

At this point, however, we should look at another consideration advanced by Malcolm. Some time before discussing the case of the ache from which he was distracted, Malcolm puts forward an argument from ordinary language to show that, in ordinary circumstances, no distinction is made between pain and awareness of pain (p.14). He considers a dentist drilling a tooth and asking his patient: 'Do you have any pain?' or 'Do you feel any pain?' or 'Are you aware of any pain?' or 'Are you conscious of any pain?' Concerning these remarks and some others he says that, as they are actually used in a dental office, they all come to exactly the same thing.

I think that in fact a number of the remarks listed by Malcolm do have the same, or pretty much the same, meaning. But let us confine ourselves to 'Do you have any pain?' and 'Are you aware of any pain?' We may concede to Malcolm that, in the situation described, the practical point of the two remarks would be the same. But that is a bad argument for thinking that the two remarks have the same meaning.

Consider the following simple situation. I am looking out of the window, and you say to me 'Is she coming down the road?' or else 'Can you see her coming down the road?' It might well be that, given the situation, the practical point of the two questions is exactly the same. Yet it is obvious that the two questions do not mean the same, and it is easy to think of situations where that distinction of meaning would be important. In the actual situation, she is coming down the road if and only if I can see that she is coming down the road, and this is mutually known to the parties to the conversation. Because of this, *and only because of this*, it does not matter which question I am asked.

Similarly, the dentist and patient can fairly assume that, in the situation the patient is in, the patient will have a pain if and only if he is aware of a pain. Nevertheless, it may still be that 'Do you have any pain?' has a different meaning from 'Are you aware of any pain?' and that in special circumstances this difference of meaning will emerge. A contemporary dentist has various techniques for dealing with a patient's pain. If he uses anaesthetic, then this presumably *stops* the pain. If he uses music, then this may in some degree *distract* the patient from his pain. But, as I argued in the case of Malcolm's aching legs, it is probable that the pain itself continues. If, as some dentists have done, he uses hypnotic techniques, it may be clear that the patient is no longer *aware* of his pain. However, as I will argue at a later point, we can still have an open mind about whether the patient still *has* his pain.

Malcolm's refusal to distinguish between being in pain and being aware of being in pain is firmly in the Wittgensteinian tradition of thought about the mind. It is interesting to observe that it represents a back-handed endorsement of the Cartesian tradition. The Wittgensteinian tradition rejects the doctrine of inner sense. As a result, for Malcolm the self-intimating and indubitable nature of inner sense can only be the compounding of error. Nevertheless, for Malcolm there is something right in the idea that if you are in pain, then you are automatically aware that you are in pain, and also that if you think that you are in pain, then you are in pain. What is right (according to him) is that there is no difference in our use of such sentences as 'He is in pain' and 'He is aware of being in pain', because the second sentence says no more than the first. The result is to set up a linguistic link between mentality and consciousness which echoes the Cartesian metaphysical link. My argument is that, whether it is conceived metaphysically or linguistically, the link does not exist.

I have argued that it is possible, and indeed likely, that Malcolm's ache went on while he was distracted from paying any attention to it. If this conclusion can be sustained in the case of unpleasant bodily sensations, such as aches, then it is all the more plausible in the case of mental phenomena which have no particular affective tone. It is no accident that Cartesians and Wittgensteinians regularly take pain as their example. It is the best ground which they have. If instead we consider ordinary perceptions, after-imagings, etc. it does not seem in the least implausible to think that they have various features which we normally take no notice of, but which we can become aware of if we attend to them.

Here, for instance, is a passage from H. H. Price:

> If one looks at a brightly luminous object, there is a characteristic series of colour-changes in the resulting after-image. How many of us had noticed these, before we read text-books of physiology or psychology? I myself, before I read William James, had never noticed that the visual size of an after-image alters very greatly if one projects it first on one's finger nail, and then on a distant wall. It is natural to say that physiologists and psychologists have *discovered* some characteristics of after-images which the rest of us had not previously noticed, but can notice now when our attention has been called to them. (Again, how many of us had noticed that we are all colour-blind in the margin of the visual field? It is natural to call this a discovery too.)[12]

I regret to say that I quoted this passage many years ago, but went on to argue that nevertheless sense-impressions could not have unnoticed features.[13] But in fact the argument of the passage seems conclusive. Against Price, I wish to deny that there are such *things* as after-images. To have an after-image is to seem to see a physical phenomenon of a certain sort: the after-image itself, I maintain, is a purely intentional object, like the thing believed in the case of a false belief. But Price seems to show clearly that the intentional object can have features of

[12]H. H. Price, Review of A. J. Ayer's *Philosophical Essays*, *Philosophical Quarterly* 5, 1955, p. 274.

[13]D. M. Armstrong, *Perception and the Physical World*, Routledge & Kegan Paul, 1961, p. 39.

which we are normally not introspectively aware, but of which we can make ourselves introspectively aware. It can seem to us that we are seeing an object of a certain sort, but we may not be introspectively aware of all that we seem to see.

So much for trying to make plausible the idea that there are mental phenomena which are, as it were, in the introspective field of view, but which are not actually objects of introspective awareness. I now go on to argue that there are mental phenomena which are not even potentially available to introspection.

The position of Freud is of great interest here, and by this I mean not simply his psychological doctrines but his scientific methodology. Freud was a scientific realist about the unconscious mind: he thought of it as a really existent thing, exerting causal power. His model was the iceberg, with the seven-eights of the berg beneath the surface representing the unconscious. Much of what is unconscious is naturally unconscious, he holds, but a certain amount of material is thrust down into the unconscious, and held there, because of its peculiarly unpleasant associations, by the mechanism of repression. The repressed material may then act like an unseen planet. The planet sets up otherwise inexplicable perturbations in the orbits of the known planets. The repressed processes perturb conscious processes similarly in an otherwise inexplicable manner.

Now I do not wish to defend any of the details of the Freudian scheme. But it seems to me likely enough that there is something in some of the Freudian hypotheses. However, the chief point I want to make is that the Freudian picture seems to be a completely intelligible picture. (It would fit well with the causal theory of the mind, to be defended in the next section. It would also fit in well with materialism, since there seems to be no particular difficulty in identifying unconscious processes with brain-processes.) If it is an intelligible picture, then current mental phenomena do not have to be self-intimating.

As a matter of fact, Malcolm seems prepared to make some concessions to a loosely Freudian point of view. At any rate, he does say, in criticism of Locke, that a jealous thought can occur to a man without his being aware of it *as a jealous thought* (p. 23). However, he does not think of the thought, as Freud would think of it, and as I would, as actually caused by occurrent jealousy, or at least a jealous impulse, an emotion or impulse which is unacknowledged and perhaps unintrospectible. I think that Malcolm's idea is rather that jealousy is a pattern occurring over time, and that the man might have the thought without being aware of its place in this temporal pattern. (If so, an isolated jealous thought might be a difficulty for him.) Hence, from the point of view of a debate with Malcolm, it may be useful to turn to another topic: that of so-called subliminal perception, perception which occurs without the perceiver being aware of it, or being able to make himself aware of it.

Actually, we have already argued for the possibility of such perceptions when we discussed the case of the sleep-walker (pp.118–20). The sleep-walker must be perceiving (it was noted that this is not the 'must' of logical necessity), but at least appears to be totally unaware of perceiving, or of any other mental state in which the sleep-walker is.

Here is an intelligible possibility. Messages are flashed upon a screen, but for so short a time that persons looking at the screen think that they are seeing nothing at all. But suppose that their conduct is significantly affected. Perhaps the message suggests that they should eat ice-cream, ice-cream is not something which they normally eat, yet they find themselves with a strong urge to eat ice-cream. If this did occur, then the natural explanation of their urge would be that they did perceive the words, although unaware of perceiving them. The verbal suggestion slipped under the guard of consciousness.

As a matter of fact nothing as straightforward (or as alarming) as this actually seems to occur. But what the case does show is that the notion of perception of which we are totally unconscious, and of which we cannot make ourselves conscious, is an intelligible supposition. Furthermore, if we look at the actual *evidence*, say that supplied by N. F. Dixon in his authoritative book on subliminal perception, we see that this evidence, although not of such a decisive nature as the ice-cream case would be if it were real, still points clearly to the reality of such perception. At the end of his book Dixon says:

> As a result of being tested in eight different contexts, subliminal stimulation has been shown to affect dreams, memory, adaption

level, conscious perception, verbal behaviour, emotional responses, drive related behaviour, and perceptual thresholds.[14]

I will finish this long discussion of the doctrine of self-intimation by considering the case which intuitively gives the greatest difficulty for my thesis, and to which Malcolm does not fail to refer (p. 9), that of severe pain. Could we be in severe bodily pain and be unaware that we were in severe pain?

The problem about such a case is that, by definition, severe pain is a sensation which arouses a strong adverse reaction. It is therefore not easily overlooked! Nevertheless, as the case of a sleep-walker shows, it is possible to be active and yet unaware of one's situation. Suppose, therefore, the following dissociated state occurred. A person goes through many of the physical routines associated with having a severe pain in the hand. The hand is nursed, favoured, and so on, and there is the restlessness associated with pain. At the same time, however, the person denies having a pain, and the denial appears to be sincere. One could imagine such a person passing the most sophisticated tests for the detection of lies. One can imagine also that when the person is faced with the paradox of the combination of the denials and the pain-behaviour, he reacts with incomprehension. Perhaps he is not even aware of the pain-behaviour.

In these circumstances, at least one reasonable way of explaining what was going on would be that the person had a severe pain but was unaware of having it.

It is clear how Malcolm would react to such a case. He would say that the speaker's denial of being in pain could not be understood. There is a tie between the language of pain and the rest of one's behaviour, he thinks, such that, if they conflict in this radical way, we cannot understand the language of pain (p.13). I allow, for the sake of argument, that Malcolm's view is a possible view to take. But my concern here is mainly to show that my view can successfully be *defended*. If there really is a faculty of inner sense, then, since it is a human faculty, it ought not to be a faculty which automatically becomes aware of everything currently in the person's mind. Indeed, it ought to be possible for any current mental phenomenon to be overlooked, including pain. I have therefore constructed a logically possible case

[14] N. F. Dixon, *Subliminal Perception: The Nature of a Controversy*, McGraw-Hill, 1971, p. 320.

which, in the context of the inner sense theory, may reasonably be construed as a case where a person is in pain but unaware of being in pain.

Nor do I have to rely solely upon a logically possible case. I can appeal to some suggestive experiments conducted by J. P. Sutcliffe.[15] His paper was published over twenty years ago, in 1961, but I understand that his results have not been overturned in the meanwhile.

Sutcliffe experimented with hypnotic techniques. He distinguished between 'credulous' and 'sceptical' views of hypnotic phenomena. A subject is successfully hypnotized and given a painful electric shock, but the hypnotist suggests to the subject that no pain is felt. The 'credulous' view is that the hypnosis actually abolishes the sensation. The subject appears to believe that this is what has happened. A typical experimenter's report on a subject runs:

> Stuporous passivity; no movement apart from hand jerk to shock. When questioned, denies feeling anything.[16]

The 'sceptical' view is that either the subject's denial is not fully sincere—he or she is more like an actor playing a role than one might think—or else that the subject *has* a pain, but, as a result of the hypnotic suggestion, believes that there is no pain. The evidence which inclined Sutcliffe to accept the 'sceptical' view is that the subject's GSR—galvanic skin response, a sensitive indicator,—is appropriate to pain rather than absence of pain.

It is clear that Sutcliffe's results are not conclusive. But they certainly suggest the possibility of pain in the absence of awareness of pain.

More recent experimental data is given by E. R. Hilgard in his book *Divided Consciousness*. It appears that in *some*, but only some, hypnotized subjects a 'hidden observer' can be contacted who has a different view of the situation from that derived from the ordinary reports of the hypnotized person. The 'hidden observer' can be contacted *via* automatic writing, automatic talking or pressing of keys. In one case where ordinarily painful stimuli were present:

[15]J. P. Sutcliffe, '"credulous" and "Sceptical" views of Hypnotic Phenomena', *Journal of Abnormal and Social Psychology* 62, 1961.

[16]Ibid., p. 194.

> The 'hidden observer' was reporting essentially normal pain while the hypnotized part of her was feeling no pain at all.[17]

Here, it seems, there really was pain, and a part of the mind was aware of this. But the 'hypnotized part' was unaware of the pain.

Against introspective infallibility

After the long discussion of self-intimation, perhaps no very extended discussion of the Cartesian doctrine of introspective infallibility is required. Some one who accepts the doctrine of inner sense, and is also a materialist, may even have some reason to think that introspective error about our current mental states will not be a conspicuous phenomenon. The causal chain between a mental phenomenon and the awareness of that mental phenomenon which it brings about is purely intracranial. As a result, it will not be subject to the interference brought about by unusual environmental conditions which can occur in sense-perception, especially a distance-sense as vision is. (Note that proprioceptive illusion is relatively rare.) Nevertheless, one would expect introspective error to occur on occasions. Certainly, it must be logically possible.

It seems enough, therefore, to discuss the most difficult case for my view: that where one seems to oneself to be in pain, but is not. And here it is simply necessary to construct a case which mirrors the case of severe pain which we are not aware of having, the case discussed at the end of the last section. Let a person assert that he has a severe pain in his hand. Let extensive tests for sincerity yield positive results. Yet let him exhibit no pain behaviour. He pays the allegedly hurting hand no special attention, and he exhibits no other pain behaviour. We can imagine that drawing his attention to the discrepancy between his assertion and his conduct produces a little perfunctory imitation of pain behaviour, but no awareness that the behaviour is merely perfunctory. Such behaviour *could* be interpreted as a case of introspective error. He thinks he is in pain, but is not in pain.

There is psychological evidence which at least suggests the possibility of introspective error even in the case of sensations of pain. The experiments performed by Sutcliffe are again relevant

[17] Hilgard, *Divided Consciousness*, p. 189.

here. Hypnotized subjects were induced, apparently sincerely, to report painful shocks, although no shocks were given.[18] It could be thought that the hypnotic suggestion actually produced pain. But this 'credulous' hypothesis is at least weakened by the finding that GSR levels were not of the height to be expected if the subject really had a sensation of pain. The 'sceptical' hypothesis that they falsely believed themselves to be in pain is a live option.

I will finish this chapter by noticing that if introspective error about current mental phenomena is allowed to exist, then a major reason for accepting the existence of *sense-data* is removed. Many philosophers in the past have thought, and a number still think, that if, say, an object looks to a certain perceiver to be red and round, then it is entailed that there is another object which *is* red and round. This object, the sense-datum, is usually conceived of as in the mind of the perceiver. Epistemological access to its redness and roundness is thought of as direct and infallible. Through it, there is non-direct knowledge of its external physical causes.

Locke and others were at least drawn to some such doctrine, but it is to be noted that it does not combine very smoothly with the doctrine of inner sense. For it means that even ordinary perception begins with an awareness of a mental object: the sense-impression. This is why Brentano, as Malcolm records, says that 'outer perception' is not really perception at all, and that mental phenomena 'are the only ones of which perception in the proper sense of the word is possible' (p. 25).

But the point to be considered here is the effect of allowing that inner sense is not infallible in its apprehension of the redness and roundness of the sense-datum. If it is not infallible, then the following situation becomes possible. A person might have a red and round sense-datum, but introspection might inform him that it was green and square. Must he then have a second-order green and square sense-datum of the original red and round sense-datum? The notion seems farcical. In any case, if second-order sense-data are admitted, then by the usual argument from parity, even a *correct* introspective apprehension of the first-order sense-datum will demand a second-order sense-datum, but this time also red and round. An infinite regress looms.

[18] Sutcliffe, '"Credulous" and "Sceptical" views', p. 193.

If, however, there is fallible introspective awareness of sense-data *without* second-order sense-data, then it becomes unclear that there is any particular reason to postulate ordinary, first-order, sense-data. If sense-data can introspectively seem to be what they are not yet this does not require second-order sense-data, then why should not physical objects perceptually seem to be what they are not *without postulating sense-data to explain this fact*? The upholder of sense-data seems to be in a dialectically very weak position.

Suppose then that one holds, as I believe one should hold, a fallibilist doctrine of inner sense. The theory of perception, that is 'outer' perception, which it is natural to combine with this is direct realism. According to this view, perception, as opposed to introspection, is always perception of something non-mental: of the physical world, including our own body. (One's own brain could be perceived by that famous philosophical instrument, the autocerebroscope. But even if materialism about the mind is true, one would not be perceiving one's mind *as* the mind.) But like introspection, perception is fallible.

3 In Defence of the Causal Theory of Mind

In *The Concept of Mind* Ryle identified the mind with (at least) bodily behaviour and dispositions to bodily behaviour. He may, or may not, have thought that there are, in addition, mental phenomena which are in some sense inner. I think that he was nearest to the truth when he was trying to explain the mental dispositionally. But it is necessary to conceive of dispositions in a way that Ryle did not conceive them: as categorical inner states of the disposed thing, states which can act as causes. And, even then, dispositions provide too simple a model for the exceedingly complex, subtle, and variegated causal roles in terms of which, I hold, mental phenomena are to be defined.

In the first part of this section I will briefly outline a causal account of dispositions. Certain features of this account will then be used to illuminate the causal theory of the mind. But the illumination will be achieved in part by *contrasting* the causal role of the mental with the much less complex causal role of dispositional states.

A causal theory of dispositions

The word 'disposition' here is a philosopher's technical term. By 'disposition' is meant such properties of material objects as brittleness, solubility and elasticity. This rubber band is elastic. If a force is suitably applied it will stretch, and will continue to stretch so long as the force continues to be applied. Remove the force, however, and the band will return to its original length. It is important to realize that what we have here are *causal* conditionals. They tell us that if the band is acted upon in certain ways, then certain effects will result. Pulling on the band is a cause. It has the effect of stretching the band. The removal of the pulling agent is a further cause, which has the effect that the band returns to its original shape.

The next question to consider is whether there is not something about the band, some virtue or nature, which is responsible for the fact that these sorts of cause acting upon the band have these sorts of effects. It is hard to doubt that the band has some such nature. However, some philosophers (e.g. H. H. Price)[1] have advanced the suggestion that it is at least not logically necessary that the band have such a nature. For myself, I do not think that Price's suggestion is ultimately intelligible. I think that the conditionals, if they are to be true, require a truth-maker in the world, something about the object in virtue of which the conditionals hold. But here I will try to steer clear of this difficult controversy. The only point needed here is that we do think that *in fact* the rubber band has some nature in virtue of which the conditionals hold of it. Contemporary science tells us that this nature is a certain feature or features of the micro-structure of the object.

It is to be noted that the band's having this nature is a cause, or part cause, of the band's stretching when the force is applied. Suitable application of force to the band plus the band's having a certain nature causes the band's stretching. Removal of the force from the stretched band plus the band's having that nature causes the band's contracting.

It seems further that we can, and sometimes do, *identify* the disposition, the elasticity of the band, with this micro-structural feature of the band. It is not linguistically improper to say that

[1] H. H. Price, *Thinking and Experience*, Hutchinson, 1953, p. 322.

the elasticity of the band *is* this micro-structural feature of the band. I do not say that we have to talk this way. We can insist, if we wish, that the elasticity of the band is nothing but the fact that, if a suitable force is applied, then . . . But we can also take the elasticity of the band to be that feature(s) of the band which is *responsible*, causally responsible, for the fact that, if a suitable force is applied, then . . .

Suppose, now, that we do think of the elasticity of the band in this more concrete way. But suppose that, at the same time, we do not yet know in any detail what this elasticity consists in. (A situation which I am in personally.) We can think of the elasticity as that property of the band, whatever it may turn out to be, which plays a certain *causal role*. (We will be referring to the property by means of a certain definite description.) The causal role is specified in the following way. The elasticity of the band is a property of the band such that *if* a suitable force is applied to the band, then the combined effect of the application of the force, together with the band's being elastic, produces a stretching of the band; and, when the external force is removed, return of the band to its original dimensions.

It is to be noticed that I have been talking about the elasticity of a particular band as opposed to elasticity generally. If we seek to identify elasticity *simpliciter* with a certain sort of micro-structure, then a complication can arise. What happens if it turns out, after investigation, that all sorts of irreducibly different micro-structures answer to the causal role of elasticity in different sorts of elastic things? One can still say that the elasticity of this band (and presumably, all elastic things of a certain sort) is a certain sort of micro-structure. But one cannot say that *elasticity* is a certain sort of micro-structure.

Suppose that the situation is of this slightly discouraging sort. Can we identify elasticity, elasticity *simpliciter*, with some non-dispositional property of elastic objects? One thing which can be done is to introduce the notion of *second-order* properties of such objects. We can say that the elastic object has *a property of having a certain property*, *viz.* that property, whatever it is for the particular object, which fills the causal role of elasticity. (Some contemporary philosophers have spoken of the property 'realizing' the causal role of elasticity in that object.)

While this seems to be a permissible way of speaking, it has what I think is a disadvantage that these higher-order properties cannot be conceived of as actual causal factors in a situation where the elastic thing is stretched. Elasticity will not be causally active. It will not *contribute* to the stretching and contracting.

What I should prefer to say is this. When we say that an object is elastic, the predicate 'elastic' applies to the object by picking out a micro-structural property of the object. This micro-structural property will play an actual causal role if the thing is stretched. But the word 'elastic' picks this property out *via* an abstract description of this property, *viz*. that here, in this object, it fills the elasticity causal role. If we say this, then we can continue to say that the elasticity of objects is a causal factor, even if the elasticity causal role is played by different sorts of structure in different sorts of elastic things.

But however this may be, it is certainly legitimate to say that the elasticity *of this band* is a certain micro-structure. As I have said before, I am not claiming anything more than that this is one legitimate way of thinking of the elasticity of the band. But if it is legitimate, then it can serve as a model for the causal theory of the mind. The elasticity of objects can be specified in terms of properties of the objects, those properties which play a certain causal role. My hypothesis is that mental phenomena—the having of purposes, perceptions, beliefs, thoughts, dreams, images, consciousness—can all be specified in a similar manner, as properties of persons, those properties which play a certain causal role.

This causal role is different for different sorts of mental item, and is a far more complex role than the role by which the elasticity of objects is specified. Nevertheless, the account given of elasticity can be used as a clue to the great labyrinth of the mental.

Different ways of developing the causal theory

The specification of the causal role of elasticity which has been given in the previous section was obtained, it seems, simply by reflecting upon the meaning of the word 'elastic', or, one might also say, unpacking the concept of elasticity. It is only necessary to think about what counts as something having the property of elasticity. One can proceed *a priori*. (Which is not to say that one can proceed with certainty.) This may be contrasted with the scientific, *a posteriori*, identification of the concrete nature, or natures, of elasticity.

This suggests that the causal theory of the mind also is an analytic theory, one concerned wholly to spell out the meaning of mental terms, or unpack the mental concepts, in a certain causal way or ways. If so, it too can be developed *a priori*. (Although, as always in such a situation, the doubt can be raised whether the concepts apply to anything actual.)

In Part II of *A Materialist Theory of the Mind* I assumed that this was the correct view of the status of the causal theory, and offered a logical analysis of all the main mental concepts, but with different sorts of mental phenomena playing different sorts of causal role. Conceptual analysis is not a very fashionable philosophical activity at the present moment, but I still think that this is a promising way to view the causal theory.

However, to take this view of the causal theory it is necessary to think that there is a reasonably sharp distinction between analytic and synthetic statements, without which a distinction between conceptual analysis and empirical investigation can hardly be made. Those who, under Quine's influence, reject the distinction between analytic and synthetic statements will not see the theory as an analysis of the mental concepts. (Nor, presumably, can they see the specification of the causal role of a thing's elasticity as anything but a primitive *theory* of elasticity. This makes their view rather implausible.) I take it, however, that such persons can at least consider the causal theory as a scientific theory. Its merits or demerits can then be discussed on this basis.

If the position is taken that the causal theory of the mind is just a scientific theory like any other, then a useful analogy is genetic theory *before* there was any identification of genes with their material bases. At that time genes were seen only as causal factors within the cell which, in conjunction with other genes, produce certain hereditary characteristics.

At this point, it may be useful to mention David Lewis' position, because it is close to my own, but not quite identical with it. Lewis, in a passage quoted by Malcolm, says that his concept of the mental is:

> . . . the concept of a system of states that together more or less

> realize the pattern of generalizations set forth in commonsense psychology . . . (p. 66)

This suggests that Lewis is thinking of generalizations like the truism that those who are injured by somebody will desire revenge. This generalization does not itself appear to be analytic. Desiring revenge for injuries done, or at least having revengeful impulses, is natural enough, indeed almost universal, but it is not part of the concept of being injured by someone. But a causal theory of mind, Lewis seems to be saying, ought to *presuppose* the truth of such a generalization.

For myself, I can see no especial objection to such generalizations being incorporated into the causal theory and so being made part of the concept of mind. But I do not think that such generalizations are the important thing. Consider, by way of contrast, the rather rough generalization that those who have simple and straightforward purposes are generally able to realize them or at least to do things conducive to realizing them. This seems to me to be genuinely analytic. It is comparable to the generalization that brittle things, when struck, generally shatter. If the latter generalization were not true, then the brittle thing could not be said to be brittle. Similarly, I think, if it were not true that those who had simple and straightforward purposes are generally able at least to do things conducive to realizing them, then those purposes could not be said to be purposes, or not said to be the purposes which they are. (If car-engines do not, in general, propel cars, then they cannot be said to be really car-engines.)

Here is another example of a causal generalization in the mental sphere which appears to be analytic. Those who believe a general proposition of the form that all F are G, and who come to believe that there is something, *a*, which is F, but who do not already believe that *a* is G, will, in general at least, be caused by these factors to believe that *a* is G. Moving from particular belief to particular belief according to this causal pattern seems to be an essential part of what it means to hold the general belief. Here we see the germ of a causal theory of *general beliefs* in terms of beliefs of a more particular sort. (A causal theory of the latter sort of belief would be still to seek.)

Consider, by way of contrast, certain points about severe

physical pain. Such pain has characteristic behavioural effects. It is presumably one of the causal generalizations set forth in common-sense psychology that severe pain causes the subject to groan, cry out and writhe. Are we to think of pain's tendency to cause these effects as part of what pain *is*? I do not think that we should. It is of the essence of pain that it is a bodily sensation, and that, in normal circumstances, this bodily sensation creates in us a most peremptory desire to be rid of it. (If the causal theory is correct, then further causal analysis will have to be given of the role of the sensation and the desire.) But the *particular* conduct towards which we are pushed by severe pain seems to be a contingent matter. Some of the behaviour serves to get attention, which may be biologically important. Some of it creates a counter-stimulus, which may in some degree distract us from the pain. But I see no reason to think that being impelled to cry, etc. is part of the *concept* of severe pain. As a result, I do not think that this characteristic result of severe pain should figure in the causal theory of pain. Lewis's formula, however, seems to cover such generalizations.

So the causal theory may be developed in various ways. I would try to present it as conceptual analysis of the mental concepts. Lewis wants to do it in terms of the generalizations of common-sense psychology. Others may think of it simply as a scientific theory about mental entities.

Mental phenomena as theoretical entities

As we have defined it, the elasticity of our rubber band is a primitive example of a *theoretical entity*, although a theoretical entity *realistically* conceived. The object exhibits certain behaviour. An underlying, unobserved, state (property) of the band is postulated which is meant to account for this behaviour, to explain why this thing, unlike most other things, exhibits this behaviour when acted upon in certain ways. Later research may, more, or less, speculatively, identify the concrete nature of this state.

If we are to carry through the analogy, then the mind and its mental states must also be thought of as (realistically conceived) theoretical entities, but theoretical entities postulated within ordinary thought and language. To embrace this analogy brings an advantage and creates an apparent difficulty. I will discuss the advantage first.

The advantage concerns the traditional epistemological problem of what good reasons we have for believing in the existence of other minds besides our own. On the traditional view, we are aware of our own mind, aware of our body, and aware of (or at least have good evidence for) various relations, principally causal, holding between our mind and our body. We observe other bodies besides our own, and observe that they are acted upon much as we are, and that they respond behaviourly much as we do. We can therefore rationally hypothesize that there are other minds standing to the other bodies in the same sort of relation which our mind stands to our body. This argument does not seem altogether a bad argument, but it does suffer from the disadvantage that our epistemological base is rather small. There is only one case, our own, to base the inference upon.

(In passing, the traditional view of the relation of mind to body has sometimes been combined with epiphenomenalism, the view that the mental is impotent to act upon the body, and in particular upon the brain. Such a view cuts away one of the important sorts of causal relation thought to hold between mind and body. So to embrace such an epiphenomenalism, as was done e.g. by T. H. Huxley,[2] is to weaken the traditional justification for believing in other minds.)

But if the mind is a theoretical postulation, in the way that the elasticity of the band is a theoretical postulation (given the realistic–scientific way that we are understanding the elasticity of the band), then there is no essential difference between the first-person case and the other-person case. In both sorts of case, an inference can be made from actually observed behaviour in the presence of stimuli. Rubber bands all behave in much the same way under the influence of certain forces. So it is reasonable to postulate a common factor, their elasticity, in all of them. Human beings all behave in fairly much the same way, especially under the influence of the same sort of stimuli. So it is reasonable to postulate the same sort of inner cause (minds), in order to explain these similarities.

As a result, the causal theory helps with the other-minds

[2] T. H. Huxley, 'On the Hypothesis that Animals are Automata, and its History', in T. H. Huxley, *Methods and Results*, Macmillan, 1894, p. 53.

problem. We have something more than our own case to argue from. But there is an obvious difficulty in treating the mind and its states as theoretical entities. Although apparently not directly aware of the minds of others, we are directly aware of our own minds. Our own mind, at least, is more than a fruitful postulation.

In answering this objection, the first thing to do is to distinguish sharply, as I have not so far done in this section, between mental states and events and processes, on the one hand, and the mind which *has* the states and in which the events and processes occur, on the other. It is plausible that we are directly (but not infallibly) aware of some of our own mental states, events and processes. But it is not very plausible that we are directly aware of our own *mind*. In the *The Problems of Philosophy* Russell suggests, somewhat tentatively, that we have direct acquaintance with our own self.[3] But Hume's view, that introspection delivers only 'particular perceptions',[4] seems more plausible.

The fact seems to be that our own mind is a theoretical postulation, even if one which appears entirely justified. In this postulation, we refer the data of introspection: perceptions, thoughts, emotions, purposes and so on: to a single organized object. (The materialist will say that it is in fact the brain, organized as the central nervous system.) It is dubious whether we would ever do this but for the teaching of those around us: we have to be taught that we have a mind. In attributing minds to others we are being doubly theoretical. We are attributing inner mental states, etc. to them which we do not directly observe. But we are also assuming that these states are organized into a mind.

So we are left with the objection that, although our own mind may be a theoretical postulation, we appear to be directly aware of some of our own mental states, events and processes. If so, how can it be said that *they* are mere creatures of theory?

The objection is just, I think. But, at the same time, there is something in the idea that *even our own introspectively observed states, events and processes* are theoretical entities. They are semi-observational, semi-theoretical entities. They are observed. But what they are observed *as* is simply as things which play

[3]Bertrand Russell, *The Problems of Philosophy*, Home University Library, 1912, ch. 5.

[4]Hume, *Treatise*, book I, part IV, sec. VI.

a causal role. We are often directly aware of our own purposes, for instance. But all that we are aware of them *as* is as states within us impelling us in certain directions, the direction being the state of affairs purposed.

Here is a model for the situation. Suppose that some glass is brittle and some is not. Suppose that somebody has the following power. Passing his fingers over pieces of glass, as a causal result of this contact he can immediately, non-inferentially, (but not of course incorrigibly!), acquire the information which pieces of glass are in the brittle state and which are not. This might serve as a grossly simplified model for introspective awareness.

As a matter of fact, tactual perception of pressure is, or can be, rather like this. Pressure is essentially a causal notion, and a sensation of pressure is a sensation as of the acting of a force upon the body. It is possible to have such a sensation without any awareness of the concrete nature of that which is acting, although there is awareness of something acting. We are simply aware of the operation of a causal factor.

The causal theory will require an account of perceptions in terms of their causal role. (I suggest that the account will involve the acquirings of selective capacities, conceived of as inner states, towards stimulus-objects acting upon the perceiver.) If a causal account of perception can be given, then it should be possible to adapt the account to fit inner perception, that is, to fit introspective awareness. All perceiving is perceiving *as*. In ordinary perception, the perceiver comes to be in informational states which enable him, if he should so purpose, to react back to aspects of the physical world in a selective manner. In introspection, therefore, the introspector comes to be in informational states concerning his own mental states. These, if the causal theory is correct, are inner states which are occupiers of complex causal roles. The information acquired about them in introspection is limited to them *qua* occupiers of these complex causal roles. This information then permits selective reaction by the person to his own mental states.

Mental phenomena have a relational essence

If the causal theory of the mind is correct, then the essence of the mental is relational. Nothing very recondite is meant by this.

The essence of being a father is largely relational, for to be a father involves having the relational property of having fathered a child. It is of the essence of the mental that it has the relational property of bringing about certain effects, or, in some cases, the property of being brought about by certain causes. It is true that some mental phenomena may not actually play a causal role, and so may lack such relational properties. In such cases, however, it will generally be of their essence that they are *apt* for having certain relational properties, because they are apt for playing a certain causal role. Some mental phenomena may not even be so apt. But it will be of their essence that they *resemble* what is so apt.

It is this fact which gives plausibility to behaviourist, and quasi-behaviourist, theories of mind. The mental, I hold, has no sort of overlap at all with behaviour. The mental is all within. Nevertheless, the mental has logical links to behaviour. For it is defined in terms of its causal role, and that causal role is spelt out by reference to behaviour (including relations to stimuli). There is a conceptual connection, a subtle form of necessary connection, holding between mental phenomena and behaviour. This connection is parallel to, although much more complex than, the conceptual connection between elasticity and the characteristic manifestations of elasticity.

It is, of course, an essential part of the causal theory that, just as the elastic nature of elastic things is not exhausted by its causal role, so the nature of the mental is not exhausted by its causal role. Mental phenomena have natures of their own, and *considered as things having that nature*, there is no necessity that they should have any particular causal powers. Malcolm makes this point an occasion for reproaching the causal theory (p. 92). According to the causal theory, he argues, it may be that my intention to paint the bathroom brings it about that I paint the bathroom. But now consider the intention as it is in its own nature: a brain-state perhaps. If the causal theory is correct, then it makes sense to say, even although it is false, that that very state might have been an intention to write a poem in which case it would not have led to my painting the bathroom. Or perhaps it might not have been an intention at all. Malcolm thinks that this is a *reductio ad absurdum of the causal theory*.

I cannot see any serious difficulty here. The general point that Malcolm makes is a consequence of the theory, a consequence which can be calmly accepted. It is true that there seems to be a difficulty in the particular case which Malcolm proposes. The difficulty arises, I think, because intentions are not to be thought of as simple causes rather like pushes and shoves. In particular, it is of their essence (*qua* intentions) that they are concept-sensitive causes, which operate by using beliefs, reasoning, recognitional capacities and so on. I cannot have the intention to paint the bathroom, or the intention to write a poem, unless I have at least the concepts of paint, bathroom, poem, etc. with all that these sophisticated concepts involve. As a result, to suppose that the cause which is in fact my intention to paint the bathroom logically could be an intention to write a poem, is to suppose a complex and messy transformation of indefinitely many elements from playing one sort of causal role to playing another. As we may put it, these mental causes are molecular, not atomic. How are we to conceive one very complex cause as if it were a quite different, but equally complex, cause? No wonder, then, that the logical possibility of the transposition seem so problematic.

Suppose, however, that we consider something simpler. It is now known that the two hemispheres of the brain are associated with different mental functions. At the same time, however, the hemisphere which does not undertake a certain task can sometimes take over from the other in the event of damage to the latter. The younger the person, the greater the chances of good transfer. If a person is actually born with damage in a hemisphere, 'transfer' of functions normally carried out by that hemisphere can be completely successful.

Suppose, then, that a process in one hemisphere plays a certain complex causal role. According to the causal theory, this causal role may constitute it a sensation, a thought, a purpose and so on. But suppose that this process does not occur in its physiologically normal hemisphere because of some earlier brain damage. It then seems an easy supposition that this sensation, say, although in fact a sensation, might not have been a sensation. For if there had been no damage to the normal hemisphere, then this process in the abnormal hemisphere could still have occurred, yet would then have failed to fill the right causal role, and so would not be a sensation. So we could say that this brain-process *is* a sensation, but yet that this brain-process *might* not have been a sensation.

Intentionality and dispositions

Malcolm briefly mentions the now very familiar view of Brentano that mental phenomena have a 'direction upon an object', 'contain an object intentionally within themselves'. As Brentano puts it, quoted by Malcolm, 'in imagination something is imagined, in judgement something is affirmed or denied, in love loved, in hate hated, in desire desired' (p. 25).

Brentano held that *every* mental phenomenon exhibits intentionality. I am sympathetic to this view, although there are apparent counter-examples (e.g. objectless depression) which need to be discussed. The question of interest at present, however, is the nature of intentionality, and whether it is *confined* to mental phenomena.

It is a rather mysterious phenomenon. Consider Wittgenstein's question 'What makes my image of him into an image of HIM?'[5] How does the image point to that particular person rather than any other person? (It might be a very vague and schematic image.) What relations must obtain between the imaging and the person imaged? And could these relations be purely physical relations?

A causal account of the mental will aim to give an account of the relation—it will certainly have to be a very complex account—in causal terms. If this can be accomplished, then we see the possibility of these causal relations being embodied in purely physical entities and systems.

However, to worry about the nature of the relations holding between say, an image and the thing imaged, is to have exposed only a part of the problem of intentionality and, in my judgement, the least worrying part. The deeper trouble is this. How can we construe the direction upon an object of the mental as *any* sort of relation between that which has the direction and the thing it is directed upon? For there to be a relation, the terms of the relation must exist. But something can be imaged, believed, loved, hated, desired, and yet that something need not exist. As

[5]Wittgenstein, *PI*, p. 177.

Wittgenstein puts it, a little brutally, a man must exist to be hanged, but can be thought of even if he does not exist.[6] How can we do this? How, in particular, can a purely material system do this? Brentano himself thought that this direction upon an object *which need not even exist* shows that the mental cannot be physical. You need a special sort of thing to carry out this special sort of task.

What I want to point out now is that a mere disposition, such as elasticity, furnishes us with a model, even if a crude and preliminary model, for this direction upon a perhaps non-existent object. In so doing, it seems to take at least some of the mystery out of intentionality.

Consider an elastic object which is in fact never stretched during the time that it remains elastic. It has a potentiality which is never manifested. The manifestation is therefore a non-existent. But may not the elastic thing be said to have 'a direction upon this object'? It points to a stretching which does not exist.

My suggestion is that a person's having a purpose, say, points to the thing purposed, in the same general way, even if in an indefinitely more complex and sophisticated manner, that an object's being elastic points to the stretching of that object. An elastic thing will stretch, if, but only if, suitable force is exerted. So it 'points' to a possible stretching. A purpose will bring about the thing purposed if, but only if, the situation is suitable. So it 'points' to the thing purposed, in the same sort of way that elasticity points to stretching.

The case of purpose is a relatively simple one, of course. It is not hard to see what constitutes the pointing character of purposes. The pointing character of, say, a false belief about some sophisticated matter in science (or philosophy!) will not be so easy to explicate. But with the analogy with dispositions, a useful start has been made. We can begin to see how a material system can point to the non-existent.

There are those philosophers who hold that where a physical object has a potentiality, such as elasticity, we need to postulate a special sort of property over and above the categorical, scientifically ascertainable, properties of the object. Such

[6]Ibid., p. 133.

special properties are sometimes spoken of as *powers*, in the case of elasticity a passive power. There are others, like myself, who would like to explain the potentialities of things simply by appealing to their categorical properties plus the laws of nature. The former view seems closer to Brentano's view of the mental phenomena: an elastic object has a power of stretching, it has 'an object within itself', just as Brentano thinks the mental state has an 'object within itself'. But notice that *whatever* view we take on this issue, the parallel with the power of the mental to point to the non-existent remains. We can take a full-blooded view of ordinary potentialities *and* of mental phenomena, postulating irreducible powers in both the physical and the mental. Or, as I would hope, we can take a less full-blooded view of both, postulating no more than categorical states and the laws of nature. But in both cases we have a breaking down of the threatened gulf between mind and matter.

If dispositionality gives us a first intimation of a direction of the mental upon something which need not exist, then we should expect there to be intermediate cases, more complex than unmanifested elasticity, less complex than failed purpose and false belief. In fact this seems to be the case. There are objects and processes in nature, and others which are the product of human design, which have this intermediate status.

Consider a thermostat in operation, set at a certain temperature. Even if the thermostat works properly, it cannot be guaranteed that the set temperature will be achieved, and, if achieved, maintained. Nevertheless, the thermostat may be said to point to this temperature. For it tends to bring about, and sustain, the temperature. It does this by measuring any discrepancy (a measurement which may on occasion be incorrect) which may obtain between the actual temperature and the set temperature, and, as a result of this, initiating action which reduces the discrepancy as far as possible.

Indeed, the operation of a thermostat serves as a first *model* for the operation of a purpose. Consider a very simple objective, such as that of remaining upright and not falling over. In order for this to be achieved, the person or animal having this objective must become aware of any motion of the body away from the vertical, and then initiate action to restore the upright situation. The perceptions involved are modelled by the thermostat's registration of temperature-discrepancy. So the registration models the perception that one is swaying. The actions initiated in order to stay upright are modelled by the thermostat's acting in a way calculated to reduce the temperature-discrepancy. Hence the thermostat acting in this way models the purpose to stay upright.

It is fairly clear that the thermostat does not have the *objective* of maintaining a certain temperature, nor does it *perceive* the discrepancy between the set temperature and the actual temperature. Nevertheless, the thermostat has a ghost-resemblance, a cartoon-resemblance, to an object which has purpose and perception. If there are causal factors within a thing which are apt for the thing acting in the sort of way that a thermostat acts, but where the factors are indefinitely more various, and indefinitely more complex, than in the case of the thermostat, we can hope to arrive at inner states of the thing which are perceptions and purposes. The pointing character of these states will be constituted by their causal powers and capacities.

At this point something interesting emerges. Despite the fact that what mental states point to need not exist, nevertheless there is something intrinsically trustworthy about perceptions and beliefs, and something intrinsically efficacious about purposes and intentions.

A thermostat has the *capacity* to keep temperature to a set level. If it does not have that capacity, then it is not really a thermostat. A thermostat which is detached from the system which it controls may still be a thermostat. But it is so only in virtue of its capacity to act as a thermostat if it is reattached. A thermostat which is broken beyond possibility of repair is a thermostat by courtesy only, in virtue of past history, much as a corpse is a person by courtesy only.

Similarly, if a purpose is to be an inner cause driving towards the bringing about of the thing purposed, then it must have a certain causal efficacy. This in turn means that the perceptions, beliefs, etc. which serve as feedback to the inner cause (*cf.* the temperature-readings of the thermostat) must be, potentially at least, reliable. A thermostatic device may be set going in situations which are too extreme for the device to operate successfully. Equally, the situation may be such that an actually held purpose cannot be carried out, perhaps because relevant perceptions are illusory or relevant beliefs false. But the thermostat, to be a thermostat, must be able to operate in normal situations. Equally, there must be a capacity to carry through our purposes in normal situations, or at least to do things conducive to realizing them, and, to this end, in these situations, perceptions must on the whole be veridical, beliefs true (and inferences correct).

A difficulty here for the analogy is that purposes can have objectives which are physically or even logically impossible. One can have the objective of constructing a perpetual motion machine or of squaring the circle. But I think it can be argued that these are logically secondary cases. Homely, down-to-earth, purposes involving those simple achievements without which more sophisticated objectives could not even be attempted, do demand a certain capacity for carrying them out, or at least advancing them. And that in turn demands a certain capacity for correct perception, belief and inference. There is a sense in which the cases where the mental is directed upon a non-existent object are the relatively special, the relatively abnormal, cases.

Further development of the causal theory

A full working out of the causal theory of the mind is an immense undertaking. It involves no less than giving an account of the different causal role of all the different sorts of mental phenomenon. No such task can be undertaken here. All I will do is make some general remarks.

An extremely important point, not duplicated in a dispositional model, is the interlocking natures of the mental phenomena. It is, in general, impossible that one sort of mental state can exist in the absence of others. One of the most important links seems to be between purpose, on the one hand, and perceptions and beliefs, on the other. That purpose demands perception and belief is not difficult to see. Purposes are information-sensitive causes. At least in simple cases, it is the perception and/or belief that the objective, the thing purposed, has been achieved which 'satisfies' the purpose, that is, brings it about that the purpose no longer drives. Similarly, on the way to the achievement of the purpose, perception of the developing situation, together with beliefs, about e.g. the relation of means to ends, clearly play an essential role.

What may not be quite so clear is that we cannot attribute perception and belief to a person unless the person is capable of purposive action. But the point does become clearer when we consider the point which so exercised Hume in his *Treatise*.[7] What is the difference between believing that a certain proposition is true, and merely entertaining a thought with the same propositional content but without any belief? It is notorious, as Hume himself acknowledged, that his own official solution in terms of degrees of vividness of image is totally unsatisfactory. But what does seem to mark off belief, or at least the central cases of belief, from mere thought is that a belief is something on which we are prepared to act. Beliefs feed in causally to our purposes, and it is their so feeding in which constitutes their being beliefs. Beliefs are not linked to any particular purpose, they are purpose-neutral. But it makes no sense to speak of beliefs which are not the beliefs of a purposive being. Similar remarks may be made about perceptions. It is of the essence of perceptions that they enable us to act (act purposively) in our environment.

The concepts of purpose, belief and perception, then, form a package-deal of concepts. They can be introduced together, or not at all. There is no parallel for this in the far less sophisticated concept of elasticity. Package-deal concepts, however, are a familiar enough phenomenon, illustrated by such cases as husband and wife, soldier and army. No husband without a wife, no wife without a husband. No soldier without armies, no armies without soldiers. What we have in the case of certain mental concepts is simply a more sophisticated and complex package-deal, a package-deal involving causal roles.

Genetic theory also provides a useful model. The genes are identified in the first place as mere causal factors, apt for the production of hereditary characteristics. But, in general, the genes do not operate separately, with each gene, in favourable circumstances, producing one particular characteristic of the organism. That the organism has a certain characteristic is explained by supposing that this result is caused by an

[7]Hume, *Treatise*, book I, part III, sec. VII, and appendix.

interacting package of genes. Some genes in the package might, in interaction with other genes, have produced a different result.

The causal role of beliefs and perceptions may be conceded, but what is the causal role of a mere thought having the same content as a belief, or an image having the same content as a perception? In the latter cases, we have mental structures which copy the belief and the perception, yet by hypothesis they do not feed into our purposes. No doubt the role of the imagination is biologically important in making our objectives, and the way in which we achieve our objectives, more sophisticated. But it is hard to believe that this causal role constitutes the *essence* of mere thoughts and mere imagings.

What seems to be needed here, as Smart long ago perceived,[8] is the notion of a mental going on which, without necessarily having a causal role itself, suitably *resembles* something which does have a causal role. The problem here is to spell out the respect of resemblance. The following analogy seems to be helpful.

The concept of poison is a causal-role concept. A poison is constituted a poison because of its capacity to kill. We can, nevertheless, form the concept of a solution, say, which contains a poison but contains it in too weak a concentration actually to poison. This solution will have a certain resemblance to a genuinely poisonous solution. Furthermore, we can imagine somebody who is able, in a direct, non-inferential way to detect this respect of resemblance. Perhaps they can put their tongue to the liquid and become aware as a result that what is present in the solution is poison in insufficient quantity actually to poison.

I do not wish to be tied down to the details of this analogy. The resemblance of thought or image to the corresponding belief or perception is no doubt quite different from the resemblance of a very weak solution of poison to a genuinely poisonous solution. But the analogy lets us see one sort of thing which would yield 'suitable resemblance' to the causal roles of beliefs and perceptions. What we are able to recognize introspectively is that thoughts and images have *some*

[8] J. J. C. Smart, 'Sensations and Brain Processes', *Philosophical Review* 68, 1959, reply to objection 4; and 'Materialism', *Journal of Philosophy* 60, 1963.

such resemblance to the corresponding beliefs and perceptions.

(Incidentally, the reliance here on mere *resemblance* to things with certain causal roles shows that the causal theory must not be conceived of simply as an 'input–output' theory. In the case of perception, an input–output account seems to be on the right track. A perception of a certain sort is a characteristic effect of a certain cause which in turn gives rise, if suitable purposes should be present, to certain selective behaviour towards the cause. But in the case of an image it seems that we must depend upon the mere *resemblance* to the corresponding perception.)

The belief that *p*, then, will be a complex structure in the mind, involving certain concepts, and feeding in, or tending to feed in, as a causal factor in any attempt to bring about any purpose to which the belief is relevant. The mere thought that *p* will lack that causal power. But the thought's structure will model the belief's structure, and, but for, say, some missing or inhibiting causal factor, would have the causal powers of the belief. If we are able introspectively to identify it as the thought that *p*, then we are recognizing this abstract and sophisticated resemblance, a causal resemblance we might call it, to the belief.

We should not despise sheer number in constituting the mentality of mental phenomena. A small band of men organized for fighting other men is not an army. As a result, the individual men, aside from some wider context of organization in which they may be embedded, are not soldiers. Make the band larger, however, and at some point we will have an army and the individual men will be soldiers. In similar fashion, we do not have *a* concept until we have *many* concepts, *a* belief until we have *many* beliefs, *a* perception until we have *many* perceptions, *a* purpose until we have *many* purposes, all organized in a single system.

It is, of course, somewhat arbitrary at what point, as the system grows, that we have an army and so have soldiers. Equally, I believe, it is somewhat arbitrary at what point, as the system grows, that we have concepts, beliefs and purposes. But this arbitrariness seems to reflect the actual way that we think about the mental. When, in the evolution of species, and in the development of the individual, does the mental come into existence? Does it not dissolve a problem to hold that an element of arbitrary decision is involved here? (Biologists have already dissolved the problem of exactly when the non-living gives place to the living in the same way.)

However, many philosophers would want to argue that no mere increase in sophistication and complexity of causal role could possibly yield the mental. They say that there is a *qualitative* difference between the mental and the non-mental which defies all piling up of quantitative complexity. I cannot prove that they are wrong, but I think it is quite plausible that a sufficient increase in complexity will be (misleadingly) experienced as a change in quality. Consider the distinction between those things which have the source of their motion largely within themselves, and those things which do not. The special nature of those things which are capable of spontaneous motion immediately strikes the senses. Indeed, it is biologically very important that it should do so, for it marks off animals from other things. But although the distinction between these things and less complexly organized things naturally strikes us as a qualitative difference, it is in fact a distinction of degree.

Topic-neutrality

If the essence of the mental is purely relational, purely a matter of what causal role is played, then the logical possibility remains that whatever in fact plays the causal role is not material. There is no logical incoherence in combining the causal theory of the mind with, say, Cartesian dualism. Mental states might be states of a spiritual substance. Perhaps it is not very easy to understand what a spiritual substance is. But it can at least be specified negatively. It can be specified as a substance which is in time, but which is not in space, and such that certain states of this substance, and certain events and processes in it, play the causal role of the mental.

It is true that the causal theory of the mind does lead naturally on to a materialist theory of the mind. For suppose that we consider all the outward physical behaviour of human beings, and other higher animals, which we take to be mind-betokening. In the light of our current knowledge, it seems quite likely that the *sole* causes of this behaviour are external physical stimuli together with internal physiological processes, in particular physiological processes in the central nervous system. But if we accept this premiss on grounds of general scientific plausibility, and also accept the causal theory of the mind, the mental must in fact be physiological.

Formally, however, the causal theory is a 'topic-neutral' theory, because it does not specify the nature of that which plays the causal role. This seems to correspond with our experience of the mental. It has often been said that the mental, as we actually experience it introspectively, is elusive, hard to pin down, as it were transparent or diaphanous. The causal theory can explain these phenomenological reports as a somewhat distorted recognition of the topic-neutral nature of our knowledge of mental phenomena. What is grasped only as something which plays a certain causal role is grasped transparently and inconclusively.

It is to be noted further that if this neutrally grasped something is in fact physical, then we can actually predict that to introspection it will seem not to be physical. The psychological mechanism involved is pleasantly illustrated in the Headless Woman illusion. To produce this illusion, a woman is placed on a suitably illuminated stage draped all in black, and a black cloth is placed over her head. In these circumstances, the spectators do not perceive the woman's head. But this real failure to see gives rise to their seeming to see that the woman lacks a head.

Now it is perfectly plain that in introspection we are not aware of mental phenomena as material states and processes. The materialist can agree with the dualist about this. But it can be predicted that, as in the case of the Headless Woman, it will *seem* that what we are aware of are *not* material states and processes. Failure to be aware of materiality will naturally be interpreted as awareness of immateriality. The introspective implausibility of materialism is therefore no argument against materialism.

Genuine duration

Following Wittgenstein in his *Zettel*, Malcolm argues that there are things which can be said to have duration, yet lack genuine duration (pp. 79–82). Looking out into the garden, I catch sight of a wren. Now I see it, then I lose sight of it. The seeing can be precisely dated and clocked. The seeing has genuine duration. Contrast this with the case of a man who has been married for a number of years. His state of being married, Malcolm asserts, lacks genuine duration. At a certain point in time the man acquired the ability to read French and has never lost it. His possession of this ability lacks genuine duration. Yesterday he formed the intention to go away today and has not changed his mind since. The intention does not have genuine duration.

The argument given by Malcolm for these contentions is that it would be pointless and nonsensical for the man to pay close attention to his situation and report, from moment to moment, whether he is at that instant married, whether he can still read French, whether he still has the intention. Malcolm draws the conclusion that such things as intentions are not mental states, enduring until the intention is abandoned, and, *a fortiori*, not material states.

If we now consider Malcolm's cases, then we notice that being married is a legal status. It depends upon the existence of certain social institutions, certain acts done by the man, a woman and others. That he is now married is therefore a complex relational property which he has. There is no particular *non-relational* state of the man which even partially constitutes his being married.

However, knowing French and having a certain intention are quite different. The knowledge and the intention must not *lapse*. And if they are not to lapse, then, in default of some inexplicable magic, must not the man continue in certain definite states (states of the brain, I would take it)? It is one of these continuing states which enables him, if he has need, to read and understand a piece of French. The other state ensures that the resolution taken yesterday has its causal effect today.

These continuing states have 'genuine duration'. If they lapsed at any point (whether or not the man was aware of their lapsing would be irrelevant), then the ability to read French would be lost, or the intention would not be carried out.

But, having got this far, it seems natural to go on and *identify* the ability and the intention with these states. After all, these states are the *cause* of the reading and understanding of pieces of French, and the *cause* of the man carrying out his intention today. Hence it seems plausible to say that the ability to speak French, and the intention, are states with genuine duration. We may conclude that it would not be nonsensical (although it might be pointless, indeed counter-productive) for somebody to report from moment to moment, that, say, his intention still stood.

The materialist identification

As we have seen, the causal theory of the mind permits and encourages, without actually entailing, the identification of the mental with the physical. But, as much contemporary discussion has brought out, the identification must not be made in too simple a way. Consider the statement that pain is firing of C-fibres in the brain. Philosophers have used this statement to illustrate materialist claims about the mind. (We need not discuss the dubious neurophysiology involved. The statement is purely for illustration.) The statement is an example of what has been called a *type-type* identity statement. The mental type *pain* is identified with the physiological type *firing of C-fibres*. A parallel would be identification of heat with the motion of molecules. The type *heat* is identified with the type *more or less violent motion of molecules*.

Now there is nothing wrong with the identification of heat with the motion of molecules. But a little thought shows that the identification of pain with the firing of C-fibres is a much more dubious proposition. Suppose, for the sake of argument, that human beings suffer pain if and only if their C-fibres are firing. How strong an argument would this be for the identification of pain with the firing of C-fibres?

A central consideration for the materialist here is that for him there is no particular reason to think that the only physically possible minds have a neurophysiological nature. For instance, a materialist will see no difficulty in the *notion* of artificial intelligence. Given a computer program of sufficient complexity and sophistication, the computer controlled by the program might exhibit intelligence, and not in any secondary or metaphorical sense of the word 'intelligence'. And if intelligence might be built into a computer, why should not mechanisms be constructed some of whose internal states could be identified as mental states, including the state of pain? But these mechanisms would not be biological objects, and so their pain would not be the firing of C-fibres.

Perhaps dogmatism should be avoided at this stage of the inquiry. It might turn out that nothing but physiological objects are capable of playing the indefinitely complex causal roles that mental phenomena must play if they are to be mental. But equally we should avoid the dogmatism of assuming in advance that this is definitely so. The mental may or may not be tied to the physiological.

But suppose that a mechanism is constructed which has mental states, and in particular has physical pain. Must there not be something in this mechanism which is equivalent to the firing of C-fibres in human beings? The answer to this, however, shows us why we should be chary about identifying pain with the firing of C-fibres. For the equivalence need only be an equivalence of causal power. If something plays the causal role of pain, in a context of other things playing the causal role of other mental phenomena, then, according to the causal theory, it is pain. And who can say, at the present stage of inquiry, just what sort of processes are capable of playing the causal role of pain? Even if pain must be something physiological, it does not have to be the same thing in different species of animal, or even in different members of the same species.

Jerry Fodor has produced a useful illustrative model here.[9] Consider the proposition that in certain sorts of engine the valve-lifters are cam-shafts. The concept of a valve-lifter is a causal concept. (Fodor would say a functional concept, but the distinction between 'causal' and 'functional', if any, can be ignored here.) A valve-lifter is whatever plays a certain (simple) causal role. A cam-shaft is a shaft of a certain sort of shape, capable, when it rotates, of carrying out the causal role of lifting valves. There is, however, no physical necessity that valves be lifted by cam- shafts. For instance, the valves might be lifted by rods. In similar fashion, even if it is true that what plays the causal role of pain in all human beings is the firing of C-fibres, it would be extremely dogmatic to assume that that which plays

[9]Jerry Fodor, *Psychological Explanation*, Random House, 1968, pp. 113–4.

the causal role of pain in minds generally must always be the firing of C- fibres.

All this will remind us of pages 139–40 above, where I discussed the possibility that what plays the causal role of elasticity in elastic things may be different for different sorts of elastic things. Different categorical properties could be responsible for the same sort of manifestation. And, indeed, that discussion was introduced with the object of preparing the way for difficulties in the identification of mental types with physiological types.

As a result of these considerations, the materialist may be tempted to retreat from a type-type identification to what has been called, not altogether happily, a *token-token* identification. This individual pain which I am having now is a certain firing of the neurons in my brain. Furthermore, any pain which anybody has is some sort of purely *physical* process. (To desert this, would be to desert materialism.) But it need not be the case that pain is the same sort of physical process in each case, unless 'same sort' is defined as 'plays the same causal role'.

But to say this seems to be to retreat too far in the face of the difficulty. It may be granted, for the reasons just discussed, that it is implausible to identify the type *pain* with a certain neurophysiological process. But what about the more narrowly conceived type: *pain in human beings*? It is quite plausible that it can be identified with some single sort of neurophysiological process. And if even that identification turns out to be too optimistic, it will presumably be possible to find still more narrowly conceived sub-types: pain in human beings of the sort X, in human beings of the sort Y, . . . and so on, where the identification can finally be effected. For, after all, the idea that the physiological nature of pain in human beings changes from occasion to occasion, or even from person to person, seems truly bizarre, although it may be a logical possibility.

But what of the type *pain*? Granted the truth of materialism, can it be said to be a physical process? It is now threatening to break up into an indefinite variety of physical processes. Yet surely it is *one* sort of thing?

We have already encountered the same potential difficulty in the case of elasticity. In the case of elasticity, one can react by taking elasticity to be a higher-order property: the property of having some property which latter property plays a certain causal role in the characteristic manifestation of elasticity. Similarly, we could say that for a being to be in pain is for that being to have the property of having some property, which latter property plays a certain causal role (the pain role). The disadvantage of speaking in this way, as I see it, is that elasticity and pain cannot then be thought of as causes.

A way out was suggested for the case of elasticity. I suggested that the predicate 'elastic' picks out what is in fact a micro-structural property of elastic things, even if different sorts of micro-structural property in different sorts of material. But it picks out this property *via* a description of this property, the description being in every case that in this object it plays the elasticity causal role.

If this is satisfactory, it can be adapted to the case of pain, or other mental types. To say that a being is in pain, a materialist will hold, is to attribute a causally efficacious property to that being, a property which happens to be a physical property. This property may be different in different sorts of beings. But the property will in each case fall under a certain description, a description used to identify it ('fix the reference'), *viz.* that in this object it plays the causal role of pain.

In this way, I hope, the materialist can allow that there need be no one sort of physical process identifiable with a mental type such as pain, and yet still identify the mental type with a physcial process.

Causation

In this section I have sketched a causal theory of the mind, and, particularly in pages 160–3, have said a little about the identification of the mental with the physical. I will finish this section by saying something about causation.

A great part of the attraction of a materialist account of the mind is that it provides a plausible unifying account of the relation of mind to matter. If the mind is physical, obeying the same laws which govern the non-mental part of nature, then we only require to postulate the properties and principles which explain the nature of the physical world to explain the realm of the mental also.

But if this unifying account is to be carried through, then mental causation cannot differ, *qua* causation, from physical causation. No doubt the term 'cause' is not a univocal term in a physical context. Striking a piece of glass may be said to be a cause of its breaking. So may the brittleness of the glass, where the brittleness is conceived of as the molecular structure of the glass. Again, the leaving of the glass in a place where it was struck can also be said to be a cause of its breaking. Here we have an external initiating condition, an internal standing condition, and a relation of the glass to its environment, all plausibly described as causes of its breaking. There is room for all these sorts of cause, and, I believe, there is also room for the notion of a *total* cause, the totality of conditions which, acting jointly, produce a certain effect.

But, I would claim, no further major distinctions require to be drawn in order to elucidate mental causation. Purposes are something like initiating causes. They actually drive the organism in certain directions. Beliefs, know-how, mental abilities and mental capacities are more like standing internal conditions. Much as the molecular structure 'enables' the striking to break the glass, so beliefs, know-how, and so on enable purposes to be achieved. There is no fundamental difference in the causal relations involved, although, of course, in the mental case the relations are far more subtle and complex.

I do not think that the causal theory of the mind need commit itself any further about the nature of this one causality, the causality which is to be found in the mental and the non-mental sphere alike. But because it is good to reveal philosophical presuppositions wherever possible, I will briefly indicate some of my current views about causality.

Cause and effect, I think, are always 'distinct existences', separate things or events, as Hume held. Furthermore, I agree with Hume that there are no logically necessary connections between distinct existences. The latter contention may perhaps be elucidated in the following way. Suppose that the cause and the effect are picked out by descriptions which abstract completely from any *relational* properties of the cause and the effect. Suppose, that is, that the descriptions confine themselves to the intrinsic, non-relational, properties of cause and effect. Under these descriptions, it will not be a logically necessary truth that the cause brings about the effect.

Despite agreeing with Hume about the contingency of the causal relation, I completely reject his view that the objective constituent of the relation is a mere matter of regular sequence. Suppose that this individual striking causes this individual breaking. According to Hume, there is nothing in the intrinsic nature of this particular sequence which makes it a causal sequence. It is constituted a causal sequence by a relational property: the fact that it is an instance of a regular sequence, that is, the fact that *other* strikings, in similar conditions, are followed by breakings. I hold, however, that where *this* striking causes *this* breaking a causal relation holds between these events which is intrinsic to the two events. This event precedes that event. This event causes that event. In neither case is the relation constituted by what happens elsewhere.

There is a logically weaker version of Hume's view according to which if this striking causes this breaking, then this particular sequence must be governed by some *law*, although a law which we who recognize the sequence as causal may yet be ignorant of.

Many contemporary philosophers hold to the following version of Hume's view. In identifying a certain particular event as a cause (this individual striking of the glass) and a further event as an effect (this breaking of the glass) one is identifying the sequence as an instance of a *law-governed* sequence. A law-governed sequence is in turn a regular or invariable sequence. Whenever those sort of antecedent conditions occur, certain consequent events always follow. But in identifying the causal sequence as a law-governed sequence, *one need not be aware of what regular sequence the sequence is an instance of.*[10]

The advantage of this view is that one often seems to be well aware that a particular sequence is a causal sequence, and it is natural to think that if it is a causal sequence it is a law-governed sequence, and yet one may be quite unable to say what the law involved is. Striking glass is not always followed by the glass

[10]See, in particular, Donald Davidson, 'Causal Relations', *Journal of Philosophy* 64, 1967.

breaking. If it does break, we assume that some law is involved, so that in just those conditions breaking will follow. But what are 'just those conditions'?

For myself, I have sympathy with this view, but have two reservations about it. First, I reject the Humean idea that a law is nothing but a regular sequence. I think that a law is a connection between properties (connection between universals). When the striking causes the breaking, both events have various properties, known and unknown. Some of these properties are connected in such a way that if an event has certain properties it is necessitated (though not *logically* necessitated) that an event with certain properties ensues. This is the law which governs the sequence. It is internal to the particular sequence, and so is not constituted merely by the fact that, on other occasions, the same sort of thing happens. The latter fact is a mere consequence of the fact that the properties are connected in this way.[11]

Second, and more threateningly, I do not think it can be shown that cause logically involves law at all, even a possibly unknown law, and however the law is conceived. This view has been argued by Elizabeth Anscombe in particular. She has claimed that it is an intelligible notion that *this* should cause *that* even although the situation is not governed by a law.[12] I find that I have to concede her the point. It seems that there could be purely singular causation.

It still seems to me, however, that it is natural to assume that cases of causality do in fact always instantiate laws. Laws, after all, give an *explanation* of why this causes that, and it is natural to look for explanations, even if it is not logically inevitable that they should be there to be found. In particular, and this is the point of interest to us here, I can see no reason to think that mental causality is specially likely to be causality without law. The world, including the mental world, exhibits a great deal of regularity (some of it only statistical). This regularity, if I am right about the nature of law, does not *constitute* the lawfulness of the world. But the regularity seems best explained by the assumption that there are laws. Having made this inference to

[11] D. M. Armstrong, *What is a Law of Nature?*, Cambridge University Press, 1983.
[12] G. E. M. Anscombe, *Causality and Determination*, Cambridge University Press, 1971.

the existence of laws, it seems natural to go on to explain *all* causality in terms of laws. (My own bet is that these laws are in every case the laws of physics.)

Given this general position (which I have outlined, rather than argued for), I can muster little sympathy for Malcolm's discussion of causation (pp. 69–75). Causation I take to be an objective relation in the world existing independently of any knowledge which we may have of it. Malcolm, however, points to *epistemological* differences between a number of cases of causation, and implies that these differences mean that we are dealing with different sorts of causality.

A woman is startled by her husband saying 'Boo', and jumps. Both know straight away that making the noise caused her to jump. Malcolm discovers that reattaching a piece of loose metal inside a lock makes the lock work again. There is a controversy between experts over the cause or causes of rising unemployment. Considering these cases, why should we assume, with Malcolm, that the nature of the causal relation is any different in the three cases? (Though it is natural to assume that the causal relations are more complex in the third case.) It is familiar experience that sudden unexpected noises cause a startle-reaction, so husband and wife both knew at once that the jumping was caused by the noise. Locks are a little more difficult, and the causes of rising unemployment may be very difficult to establish. But what have these epistemological differences got to do with the nature of causality?

Malcolm also considers the case of a chess player who makes a strange move with the object of disconcerting his opponent. We might refer to this reason as the *cause* of his making the strange move, but Malcolm wants to deny that the reason is a cause of the move in the same sense that, say, the noise caused startled jumping.

The man desired to disconcert his opponent. It is rather common to speak of his reason as that which he desired to do: *viz*. disconcert his opponent. If that is what is called his reason in this case, then the reason is not strictly a cause. Nevertheless, even if *disconcerting his opponent* or *that his opponent be disconcerted* is not a cause, it seems at least reasonable to say that *his desire that his opponent be disconcerted* is a cause. This desire, in conjunction with the belief that this move would disconcert his opponent, brought about, that is, caused, the action of moving the piece in the particular way that desires and beliefs in conjunction do cause actions. That, at any rate, is what an upholder of the causal theory of mind would argue, and I do not see that Malcolm has given any reason here for rejecting the causal theory.

Malcolm does also mention the case where the player's plan succeeds and his opponent is disconcerted. He claims that this is a case where cause and effect cannot be conceived as 'distinct existences'. The opponent's being disconcerted, the effect, can only be understood as being disconcerted *at* or *by* the other player's strange move.

Perhaps part of the trouble here is that Malcolm has not specified the causal sequence in as full a way as is available. It is the opponent's *belief* that the other player has made the move he has made (a belief caused in the opponent by the actual making of the move), together with the belief that, given the position, this is an unusual and inappropropriate move to make, which caused the opponent to be disconcerted. Granted that, it must also be granted Malcolm that the being disconcerted is a mental state with an intentional object, *viz.* that which disconcerted the opponent: *the other player's move*.

But to raise this problem is only to raise the problem of intentionality, discussed in pages 149-53. Malcolm takes a tangled case. We have seen, however, that in less tangled cases the problem does not seem insoluble. The problem of 'distinct existences' was discussed in pages 147–8. I argued there that in the case of purposes we can conceive of them as 'distinct existences' from the thing which they bring about. The purpose, characterized as a purpose, is described in terms of what it tends to bring about. But it is possible to characterize it independently, for instance, if one accepts materialism, as a brain-state or process. It is then clear that it *is* a distinct existence from the coming to be of the state of affairs purposed.

Now, given that this approach is successful in the purpose case, then it should be possible to apply the approach to more complex cases. In Malcolm's case the opponent acquires certain complex beliefs, and is further disconcerted by what he believes to have happened. To give an account of the intentionality of all these mental states in causal terms would be a complex matter, which I can hardly undertake here. (For belief see Part I of my *Belief, Truth and Knowledge*,[13] but even that discussion does not do anything like full justice to the complexity of the topic.) But it seems a promising bet that the task can be carried out in such a way that one could then go on to identify the states involved as states of the brain. At that point it will also become clear that the causes and effects involved are, after all, distinct existences.

4 Qualities

What are the major difficulties which face a causal theory of the mind? One, certainly, is that of giving a purely causal account of intentionality. I have said something about some aspects of this problem in the previous section. But many philosophers have argued that introspection reveals that mental phenomena involve a qualitative component not to be captured by any analysis in terms of mere causal role. In this section, the objection from quality will be considered.

Malcolm also criticizes the notion of introspectible inner qualities. But I am in fact much more sympathetic to the idea that there are such qualities, than I am to Malcolm's criticism. Malcolm is hostile to the whole idea of introspection and qualities perceived only by the introspector. I have no such hostility. All I shall be arguing is that *in fact* there are no such qualities.

Perceptual qualities

If we consider belief, thought, purpose and cognate mental phenomena, then, as we have noted, they have a certain introspective elusiveness or transparency. It is at least a possible explanation of this emptiness that we are aware of these things purely in terms of their causal role. But, it is argued, when we are aware of our own perceiving, imaging, the having of bodily sensations and emotions, we are aware of a rich range

[13] D. M. Armstrong, *Belief, Truth and Knowledge*, Cambridge University Press, 1973.

of qualities. These qualities bestow causal powers, no doubt. But they are not to be reduced to mere causal powers, and awareness of them is more than an awareness of mere causal powers.

In this section I will argue that *perception*, as we experience it introspectively, is entirely qualityless. The only qualities involved are qualities, not of mental phenomena, but of the physical things perceived. In particular, I maintain that the so-called secondary qualities: colour, sound, taste, smell, heat and cold: qualities which have often been thought to be inner qualities, are in fact qualities of objective physical phenomena.

Philosophical discussions of perception instinctively gravitate towards colour, so let us yield to temptation and begin with it. I look out of my window and see the green leaves of a vine. Colour, once we give it our attention, is a very striking and arresting phenomenon. But it does not seem to be a mental phenomenon. If we consult perception, then its verdict is clear. The green colour is a property of the vine leaves. Nor is it a property which involves any relations that the leaves have to any other object, whether objects in the field of view or the perceiver. It is an intrinsic property of the leaves. In what follows I will hold fast to this perceptual deliverance. It is not an indubitable given to be respected at all costs. There are no indubitable givens. But it is a plain deliverance of perception, and it is plausible that a true theory of perception will uphold this deliverance.

An opponent of the causal theory of the mind, however, has the following interest in holding the deliverance false. Suppose it can be shown that the perceived greenness does not really qualify the leaves, but instead qualifies something mental. This will have the consequence that when we are aware of greenness we are *introspectively* aware of it. But the things, or properties of things, we are introspectively aware of are, presumably, all mental. Yet the concept of greenness, it seems perfectly clear, is not a causal concept. So there will be something mental which falsifies the causal theory of the mind.

There are two main lines of argument by which philosophers and others have sought to show that the greenness is not a quality of the leaves. The first turns upon the fact of perceptual illusion (and mental imagery), the second upon questions of scientific plausibility. I think that neither succeeds. But the second creates rather more worry for the causal theory, and requires relatively heroic measures to meet it.

For the present, however, let us consider perceptual illusion and imagery. It is possible for something to look green, although in fact the thing is not green at all. It is possible to have an hallucination, say of a green snake, although there is nothing in the physical world which corresponds, even inaccurately, to the hallucination. One can dream of, or form an image of, something green. In such cases, it is argued, greenness is present. At the same time, by hypothesis, it is not some ordinary physical object or surface which has the quality. It is then plausible to think that it is something mental which is green.

Once one has gone this far, it proves difficult to maintain that *anything* except mental things are green. The greenness of vine-leaves is dismissed as a mere *façon de parler*. Vine-leaves are 'green' because they have the power to create in us mental phenomena which have the actual quality of greenness.

What are these mental phenomena? Here there is division. According to some philosophers, when, as we ordinarily say, we perceive a green leaf, perceiving it as green, it would be truer to say that it is the mental act of perceiving which has the green quality. We perceive greenly. According to another, historically more popular, view, we ought to postulate special mental objects, often called sense-data. It is sense-data which are the direct or immediate objects of perception and it is they which are the bearers of the green quality. (This second view seems very much preferable *phenomenologically*.)

But wherever we locate the quality of greenness in the mind, its presence there will falsify a purely causal theory of the mind.

In my view, what the causal theorist should do is to deny the first step in this argument. He should deny that when something physical looks green to somebody, but is not green, or where somebody images something green, then the sensory quality of greenness is present. The causal theorist can admit that there is a sense in which sensory illusion, or the having of such images, involves something green. But (a) the something is an ordinary physical something; and (b) it is a merely *intentional*, not a real, object.

Suppose that somebody believes that something which they

cannot at the time see is green. This belief is compatible with the object existing, but not being green, or even with the object not existing at all. A natural view to take is that the belief is a structural state, a structure corresponding to the concepts involved in the belief. The state has an intentional object: the object's being green, or, if you like, the proposition that the object is green. It is the program of the causal theory to give an account of this intentionality purely in terms of the causal role played by the mental constituents of the state and the internal organization of these constituents.

The obvious suggestion for the causal theory is to give a similar account of perception. A perception of something green will involve a green-sensitive element, that is to say, something which, in a normal environment, is characteristically brought into existence by green things, and which in turn permits the perceiver, if he should so desire, to discriminate by his behaviour the object from things which are not green. In the case of sensory illusion, a thing which is not green, but really is perceived, brings the green-sensitive element into existence in the mind. In hallucination, the 'thing perceived' has a merely intentional existence, and the green-sensitive element comes to be as a result of other causes. In having a mental image of something green, there is something in the mind which resembles a perceptual green-sensitive element, but which lacks the causal powers associated with genuine green-sensitive elements.

A particular word about mental images before passing them by. An appealing thought is that having a perception (say that something is green) stands to having the corresponding image as holding a belief stands to merely entertaining the corresponding thought. It was suggested on pages 155-6 above that, when a proposition is entertained without belief, a belief-like structure comes to be in the mind, but that the structure lacks the power to influence action which belief has. In being introspectively aware that we are having such a thought we are aware (i) of the resemblance to the corresponding belief-structure; yet also (ii) that *this* structure does not have the power to feed into our practical reasonings in the way that beliefs do. So, it may also be suggested, the having of an image is the having of a perception- like state, but where the 'perception' lacks the power to influence action. (Both thoughts and images are, or often are, under the control of the will in the way that beliefs and perceptions are not. So they are not suited to be a *guide* to the will, as beliefs and perceptions are.)

Whether we should actually *reduce* perceptions to a certain species of acquirings of beliefs (and so the having of images to a species of entertaining thoughts) is a further question. I incline to favour such a reduction, but it is controversial among philosophers of perception. What is perhaps a little less controversial, because weaker, and so may secure wider agreement, is this: perceptions are propositional in structure. To say this is not, of course, to say that they are in any way linguistic. But they do encode information (and misinformation) about the current state of the perceiver's physical environment, his body, and the relations of the body to his environment. Whether this information acquired constitutes a set of beliefs (fading with extreme rapidity for the most part), or whether we should think of it as sub-doxastic, beneath belief, but capable of *causing* fully-fledged belief, is perhaps a minor matter which we can afford to bracket here.

Whether it be belief or not, this information (misinformation) gives the perceiver the capacity to react back upon the current environment and his own body if he should purpose so to react. Among the phenomena which support this informational account is the indeterminacy of perception. I can perceive that the vine-leaf is green, and even see what shade of green it is although I have no name for the shade. I can perceive its shape and size, and distance from me. But I cannot perceive its *exact* shade, its *exact* shape, size and distance. The perception is indeterminate in these respects. This naturally suggests the indeterminacy of most (all?) information.

(Those who postulate sense-data to explain the perception of the vine-leaf must either say that sense-data are indeterminate in nature, making sense-data paradoxical entities indeed, or else say that sense-data are perfectly determinate but that our apprehension of them is indeterminate. Against the latter way out the following may be said. Ordinary vine-leaves can be further investigated in individual cases and their colour, shape, size and so forth more exactly determined. Nobody has ever suggested how this may be done with sense-data. The hypothesis of sense-data is therefore at a methodological disadvantage.)

What I have given is a very brief sketch of a theory of perception compatible with the causal theory of the mind. If it is along the right track, then there is no call to treat illusory or imaged sensible qualities, and in particular colour, sound, taste, smell, heat or cold, as actual qualities of actual entities. The case of colour was taken, but the argument goes through in the case of all the qualities. What we are dealing with here is misinformation, or, in the case of imaging, something like the entertaining of fantasies. The illusory qualities do not qualify anything.

From this perspective, therefore, sensory illusion and the having of images give us no reason to think that perceived or imaged colour, sound, taste, smell and so on qualify anything mental. We can be introspectively aware of the having of a perception (or image) of something green, or of a perception (or image) of something red, and introspectively aware of the difference between the two perceptions (or images). But why should we think that the introspective awareness is an awareness of two different *qualities*? Redness and greenness are two different qualities, of course, but they are qualities of things without. A green-sensitive element within need not be green, nor is it introspected as something green, nor indeed is it introspected as having any quality at all. It is introspected simply as something having sophisticated causal relationships to green things. A purely causal theory of the nature of perceptions (and images) can so far be sustained.

This result, however, is challenged by a more formidable set of arguments, drawn from scientific considerations, which seem to show that the secondary qualities cannot qualify external things. The importance of these arguments makes further comments desirable.

Physics and the secondary qualities

The traditional distinction between the primary and the secondary qualities became prominent in the sixteenth and seventeenth centuries with the rise of modern science. It is a product of that rise. Despite the attempt by Locke to give the distinction an *a priori* basis,[1] its foundation appears to be *a posteriori*. To thinkers such as Galileo, Newton and Boyle it seemed that some of the perceived qualities of objects played no role in determining the behaviour of the objects. That behaviour appeared to depend upon the shape, size, position, state of motion and mass of the objects. But such qualities as colour, sound, taste and smell seemed to play no role in 'the executive order of nature'. Boyle called the two groups of qualities the primary and the secondary qualities respectively. Among philosophers, at least, the terms have stuck.

The membership of the two groups has not always remained constant. The reason for this is that the distinction is always drawn relative to the science, in particular the physics, of the day. The situation may be illustrated by considering the case of degrees of heat. At a certain stage, many natural philosophers thought that degrees of heat in objects were different proportional quantitites of *caloric fluid*. Quantity of caloric was then for them a primary quality of objects. But this theory was superseded by the view that degrees of heat in the object are nothing but more or less violent motion of the constituent parts (in particular, the molecules) of the hot thing. So degrees of heat in objects can be explained without postulating any primary quality of the objects except motion of parts.

Apparently successful reductions of this sort effect an intellectual economy in the theory of the physical world, and to that extent are evidently to be desired. But they also create a problem. The perceived secondary qualities seem certainly to exist. It is difficult to treat them as being, in Berkeley's phrase, 'a false imaginary glare'.[2] But if they exist, then they must qualify something. But what do they qualify? Nothing

[1] John Locke, *Essay*, book II, ch. VIII, sec. 9.

[2] George Berkeley, *Second Dialogue*, in *Berkeley's Philosophical Writings*, edited by D. M. Armstrong, Collier-Macmillan, 1965, p. 174.

physical apparently. So they qualify something mental. They must be swept into the philosophical dustbin of the mind. They must qualify sense-data ('ideas'), or perceivings, or something of that kind.

In this way we are led to the view that the truth-conditions for the statement that the vine-leaf is green is that it *looks* green to normal human perceivers, where 'looks green' is spelt out in such a way that the sensible quality of greenness qualifies something in the mind of the perceiver. And so for the rest of the secondary qualities.

This line of argument, rooted as it is in *a posteriori* scientific considerations, strikes me as having much greater force against the causal theory of the mind than the argument from sensory illusion and mental images. How should the causal theorist answer it? I will briefly consider two answers which I think are unsuccessful, and then propose another solution.

It has sometimes been suggested that we should accept the point made by physicists that such qualities as colour bestow no causal power upon the physical objects which have them, yet at the same time leave the qualities where they look to be: qualifying external objects. They are causally idle properties, getting a free ride from the objects which they qualify. This is a direct realist, non-reductive, but epiphenomenalist, theory of the secondary qualities.

This view leads to various difficulties. There are difficulties in conceiving how qualities such as colour attach to physical objects conceived of as modern physics conceives the objects. But the most serious difficulty for this view of the secondary qualities is its epiphenomenalism. If the qualities bestow no causal power, then our perceptions would be exactly the same whether or not these qualities existed. The green leaves stimulate my eyes and cause me to perceive them, but they do this because of the light-waves emitted from the object. It is the light-waves which take executive action. They in turn depend solely upon primary properties of the surfaces of the leaves. Why should we think that the leaves have any other properties? Our perceptions would be exactly the same whether or not the causally idle external greenness was there.

So much for a first attempt to escape the dilemma posed by the physicist's assault upon the secondary qualities. Can we defend the objectivity of the secondary qualities by advancing a causal theory of these properties? To do this, we must first say that for a physical surface to be green is for it to have the power to furnish green sensations to normal perceivers in standard conditions. This was more or less the course favoured by Locke.[3] Powers, however, are very close to dispositions. Indeed, we could also speak of a disposition to furnish green sensations to normal perceivers in standard conditions. As a result we can go on to give the same account of the power which I earlier gave of dispositions (pp. 137–40). The power can be identified with the property or properties of the object in virtue of which the object exerts its power. This in turn permits an identification of the colour of the object with whatever physically respectable properties of the object are the causes of the sensation. In this way, sounds can be identified with sound-waves, heat with motion of molecules, and so on.

The bulge in the carpet is here pushed down in a very satisfactory manner. The problem, however, is that it immediately appears elsewhere. What account is to be given of sensations of green? This was no special problem for Locke, for whom immediately or directly apprehended greenness was a property of 'ideas' in the mind, 'ideas' being the forerunners of the modern sense-data. But what is an upholder of the causal theory of the mind to say about sensations of green, after adopting a Lockean theory of objective greenness? Only, apparently, that a sensation of green is that which is apt to be produced in us by a green surface. We thus have a tight circle of two concepts, each defined in terms of the other. This would not be a vicious circularity if we have an independent way of introducing the two together. But what is this way? The Lockean approach gives us no help.

It has been suggested to me that the difficulty here lies in developing the causal theory as a *conceptual* theory. Suppose, instead, that we develop the causal theory as a scientific theory. (See pp. 140–3 above, concerning different ways of

[3]Locke, *Essay*, book II, ch. VIII, sec. 10.

developing the causal theory.) We then define greenness as that which is apt for the production of sensations of greenness. And for a causal theory of sensations of greenness we simply pin our hopes on future science.

This policy of the promissory note would perhaps be justified if we could see some possible way in which a non-circular causal theory of sensations of greenness could then be developed. Since I at least am unable to see this, I will try to develop an alternative.

I believe that this alternative should still hold to the idea of reducing the secondary qualities to scientifically respectable primary qualities: reflectance properties of surfaces, properties of air-waves, motions of constituent molecules, and so on. This reduction has two great advantages. The first is a phenomenological advantage. The secondary qualities can be located where they appear to be. It is the surface of the leaf which is green, sounds can fill a room, smells hang around in them, tastes can inhere in the tasty body, water can be hot or cold, just as perception delivers. The second advantage is that the qualities can be treated as causally active, as bestowing active and passive power upon the objects which possess them. As a result, there is no epistemological problem how it is that we come to be aware of them. Coloured objects act upon us *in virtue of their colour*, and create in us a perception of their colour.

At the same time, however, a causal theory of the secondary qualities seems unsatisfactory, for the reasons which have just been given. We want to see how we can get a grip upon the secondary qualities quite independently of our grip upon sensations of such qualities. I propose, to this end, what might be tagged a *Gestalt* theory. When in perception we are aware of the colour, sound, smell, taste, etc., of the physical things, then the qualities which we are aware of are complexes of physical properties. The perceived secondary qualities are primary qualities! But we are aware of them in a unified, *Gestalt*, manner, a manner which fails to reveal the primary nature of these properties.

As a preliminary model, consider that it is often possible to recognize, say, a certain complex shape without being able to analyse the shape and give the shape- formula. (The shapes might be shown for only a very short time.) A certain shape is instantaneously recognized as the same shape again, but the recognition-process does not appear, at least, to involve recognizing the shape-formula. Contemporary psychologists have spoken of the way that a stimulus from without is fitted to a template within. There could be knowledge that the right fit had been achieved, without knowing just what shape the thing perceived has which makes it fit the template rightly. Might not the secondary qualities be objective complexes of primary qualities which are recognized in this primitive manner?

However, there is a difficulty in this model. When complex shapes are shown to us quickly, and we then prove able to recognize that it is the same, or not the same, shape which is shown again, yet are unable to give the shape-formula, still we do recognize that it is *shapes* that we are dealing with. We attribute a shape to the object perceived, even if indeterminately. But when a surface looks green to me, I am not attributing the corresponding primary quality or, indeed, any primary quality. Yet I am attributing some property to it. So am I not attributing a property to the surface which is *different* from the corresponding primary quality?

The only way, I think, that this difficulty can be met is to have recourse to the topic-neutral manœuvre. Why should it not be that we attribute a property *of some sort* to the surface of the leaf, a property detected by the eyes, but without any specification of the sort, thus leaving it open that the property is in fact a primary property? It may be added to this, along the lines of the Headless Woman illusion, that our visual inability to pick out that the quality involved is a primary quality inevitably generates the illusion that it is not a primary quality. Again, our inability to pick out that the property is a structured property inevitably generates the illusion that it is not a structured property.

I believe that this suggestion is along the right lines. But, at first sight, it faces an enormous phenomenological difficulty. Earlier, I spoke of the phenomenological advantages of locating the colour where it appears to be: on the surface of a leaf. A topic-neutral account of the *mental* is reasonably plausible, at any rate if the sensible qualities are extruded from the mind. But the sensible qualities themselves are the paradigms of concrete perceived qualities. How can a sub-class of these qualities, the secondary qualities, be treated as qualities we know not what, later identified, as a result of scientific considerations, with the primary qualities?

My suggestion is that the illusion of concrete secondary quality is created in the following way. Phenomenologically, the secondary qualities lack structure, they do not appear to have any 'grain' as Wilfrid Sellars puts it. Nevertheless, they have a huge multitude of systematic resemblances and differences *to each other*. Each secondary quality has a position in complex dimensional arrays of qualities. The phenomenon is most clearly evident in the relations of the colours to each other, but is present in the case of each sense. There even seem to be classificatory relationships which cross the senses: for instance, red is to hot as blue is to cold. It is this immensely complex network of perceived resemblances and differences which largely creates the impression that our acquaintance with the secondary qualities is acquaintance with definite qualities which are other than the primary properties. (I say 'largely', because the failure to perceive the identity with complexes of primary properties, and the failure to perceive any 'grain' in the secondary qualities also have their influence, in the style of the Headless Woman.)

Resemblance is an internal relation: it depends upon the nature of the things which resemble each other. (Some philosophers had denied this. I believe that they are wrong, but it is at least profoundly natural to treat resemblance as an internal relation.) We can perceive resemblance without perceiving the respect of resemblance, and if my view of the secondary qualities is correct that is what occurs in their case. The immensely complex dimensional classification of the secondary qualities, with all its degrees of resemblance, is a matter of perception of resemblances without grasping the basis of the resemblance in the primary qualities. (A basis which scientific investigation increasingly uncovers.) However, given our instinctive taking of resemblance to be an internal relation, a mere perception of resemblance suffices to generate the illusion that we have a concrete acquaintance with the qualities which sustain the resemblance. *A perception of the internal relation of resemblance generates the illusion of a perception of intrinsic quality*.

It is worth noticing that our awareness of the resemblances and differences of colour are sharper and clearer than in the case of the other secondary qualities. (We see straightaway that orange is between red and yellow.) And it is colour which gives us the strongest impression of acquaintance with the concrete nature of the quality involved.

I conclude, then, that it is possible to uphold an identification of the secondary qualities with primary qualities. If so, there is still no reason to bring the secondary qualities within the mind, or give an analysis of secondary qualities in terms of sensations of such qualities.

Two final points before concluding. First, as in the case of the identification of the mental with the physical, there may be problems in a straightforward identification of secondary qualities with primary qualities. It has been argued that the primary qualities correlated with the one shade of colour form an irreducibly disjunctive set. (This situation does not hold for all ranges of the secondary qualities, but it may for colour.) There seems to be no *a priori* reason why this should not be so. The human perceptual system might classify together surfaces which a mature science would not treat as having anything significant in common. If this were so, and provided the evidence did not impugn the whole project of the reduction of secondary qualities to primary qualities, then the colour when possessed by *this* surface, and no doubt the same colour when possessed by surfaces of the same sort, would be identified with a certain primary property. But that colour *simpliciter* could not be identified with that primary property.

If all this were so, I think that one could simply accept it as a complication, but not as a refutation, of the account of the secondary qualities which I have proposed.

The second point is that one might wonder whether primary properties are in any more secure position than secondary qualities. Might it not turn out that they, too, are only known in terms of their resemblances and differences to the other primary qualities? If so, a topic-neutral analysis will have to be given of them, too. We will not be acquainted with intrinsic quality at any point.

Here again I would simply allow this speculation. Contemporary physics suggests that we should give an account of colour, sound, taste, smell, heat and cold in terms of the 'executive' primary properties. But who knows if the latter are fundamental? (Why should middle-sized creatures like ourselves be in perceptual touch with the fundamental properties of the world, if there are any?) A deeper physics might give an account of the current list of primary qualities in terms of properties which we can neither perceive nor image.

Bodily sensations and emotions

But what of the qualities which appear to be apprehended when we have such bodily sensations as pains, itches, tickles, tingles? Are not these qualities, qualities of sensations? And since sensations are mental, we appear to have here a falsifier of a purely causal theory of the mind.

In response to this argument I shall argue that the qualities involved qualify portions of our body. They can first be assimilated to the secondary qualities, and then identified with the primary qualities in the way suggested in the previous section.

We say that we have a pain in the hand. The *sensation* of pain can hardly be in the hand, for sensations are in minds and the hand is not part of the mind. A brain kept functioning without its body might suffer pain if stimulated in certain ways. It is clear, then, that the statement that the pain is in the hand requires some analysis. My suggestion is that what we have here is a form of perception: bodily perception or proprioception. When somebody is said to have a pain in the hand it feels to them (this being the 'feels' of bodily perception) that a disturbance is taking place in the hand. This perception can be illusory perception, perhaps a case of what psychologists and physiologists call referred pain. If the pain in the hand is referred, the bodily disturbance causing the sensation is elsewhere, but is felt to be occurring in the hand. This is like other illusions of spatial displacement found in connection with other sense modalities. For simplicity's sake, however, let us suppose the perception is veridical. The disturbance which causes a perception of disturbance in the hand is actually in the hand.

The question then is: what is the disturbance in the hand perceived *as*? It is here, I suggest, that the parallel with the secondary qualities is to be drawn. The disturbance is perceived as having non-relational qualities which apparently go beyond the primary qualities, which apparently go also beyond the orthodox list of secondary qualities, but which may be fairly compared with the secondary qualities. Different sorts of pain will involve different qualities at the 'places where it hurts' (quantitative features will be involved also) and still further qualities will be associated with the place of itches, tickles, tingles.

On this view, besides its ordinary perceived primary and secondary qualities, the body is perceived as a field of apparently extra qualities, variously located and enduring for a greater or lesser time. There is, however, no independent place for these extra qualities in a materialist world view. The materialist, therefore, will go on to identify the qualities with various sorts of biologically important disturbances in different parts of the body, disturbances whose nature involves nothing more than the primary qualities. But, like the orthodox secondary qualities, these extra qualities do not *present* themselves as complex primary qualities. A special feature of perceptions of the extra qualities, whether veridical or illusory, is that they tend to evoke certain mental reactions (desire to scratch in the case of itch, desire for the perception to cease in the case of pain, and so on).

An upholder of this view will have to accept two consequences of it. First, that the qualities could qualify the perceiver's body, but not be perceived. Second, the logical possibility of intersubjective perception of these qualities. It does not seem difficult to live with these consequences.

Suppose that somebody has a pain in the hand, that the pain is not a referred pain, but that the pain-nerve is subsequently severed. Since pain is a sensation, the pain presumably ceases. But a pain is a perception of a bodily disturbance having a certain sort of non-relational quality, a perception which evokes distress. Possession of that quality by the hand is simply part of what it is to have the pain. So, although there is no 'pain in the hand', it is still logically possible that the hand continues to have the quality which was perceived in the hand. If we further accept the reduction of the quality to primary properties, this becomes no more than the easily envisaged possibility that the physical disturbance in the hand should continue, although it no longer affects the central nervous system.

Concerning intersubjective perception, the problem is that in bodily perception we are each confined to our own bodies. But if we fantasize that a number of people are suitably wired up to the hand of just one of them, then we can see how direct intersubjective checks upon the qualities involved could be made. Indeed, we could imagine that in these circumstances a more refined discrimination of the qualities would become possible, not to mention, in the case of pain and itch, intersubjective checks on the magnitude of the disturbance.

If this is correct, then the causal theory of the mind does not require to be modified to accommodate bodily sensations. There is still no need to postulate qualities of inner (mental) processes.

The other mental phenomena which may seem to call for the postulation of special mental qualities are the emotions. It is natural to speak of the emotions as *felt*, and this may suggest that being aware of our own emotions involves being aware of certain qualities.

It is, however, clear that different emotions are associated with different patterns of bodily sensation. The James–Lange theory even identified the emotions with these patterns of sensation.[4] That cannot be correct: besides sensations, emotions involve beliefs and attitudes in complex causal patterns, patterns which differ from one sort of emotion to another. But it is plausible that the special *feel* of anger, shame, embarrassment, and so on, comes wholly from the idiosyncratic pattern of bodily sensation associated with that emotion. If that is so, and if the account given of bodily sensation in this section is correct, then the emotions do not

[4] W. James, *The Principles of Psychology*, Dover, 1950, vol. II, ch. XXV.

provide a counter-example to the causal theory of the mind. The mind as it is introspectively experienced remains a mere field of causally defined factors.

In concluding, however, it is right to note that the argument has not *proved* that we are not introspectively aware of qualities in the case of the emotions, or in the case of any other sorts of mental state. It remains possible that the causal theory does not fully capture the mind as it presents itself to itself.

Suppose then, what I have been engaged in arguing against, that there *are* introspective qualities. The causal theory will not then be the whole truth about the mind. How important would this be? The matter seems to turn upon whether the account of the secondary qualities suggested in this chapter is or is not satisfactory. Suppose that it is satisfactory, and that the secondary qualities are primary qualities not apprehended as such. Why should we not then give the same account of the 'introspective secondary qualities'? A mental process is a brain-process having nothing but physical properties. When this brain-process is introspectively perceived as having certain intrinsic properties, introspectively unanalysable (and causally unanalysable!), these properties may nevertheless still *be* complex physical properties. So the existence of introspectively apprehended inner qualities might show that the causal theory of the mind was incomplete, *but need not affect materialism*.

Suppose, however, that the secondary qualities are irreducible. The same will have to be said, presumably, about the qualities associated with bodily sensations and perhaps other mental processes, such as the emotions. The body and the mind will then have associated with them a range of qualities not to be found in ordinary physical objects. There will then be a case for thinking that body and mind are special sorts of objects, set apart from ordinary material things. For a materialist, then, much turns upon whether he can give a satisfactory materialist theory of the secondary qualities.

What is it like to be a bat?

It is clear that a bat, which has very different senses from ours, will have very different perceptions from ours. It is not clear whether a bat has different experiences from us because it is not clear whether bats have experiences at all. To speak of 'experiences' seems to imply introspective awareness of perceptions, sensations, and other mental states, but it is not clear that bats have any introspective awareness. If they do have introspective awareness, then it must be of a very primitive sort. A bat's mental life is probably more like the mental life of a human sleep-walker than a conscious human being. However, bats certainly have perceptions, and these perceptions will be very different from ours.

One question is whether the causal theory of the mind, and the physicalist account given of the secondary qualities in this section, can do justice to this difference.

I do not see why not. Perceptions, according to the causal theory, are inner states which permit selective behaviour towards features of the environment. Their special 'quality' is intentional only: they point towards the features which they are perceptions of. Is the bat offered as a counter-example to this view? If so, I cannot see it adds anything much to arguments for sense-data, or for inner mental qualities generally. All that we need to admit is that the intentional objects of bats' perceptions are utterly different from our own.

The conspicuous thing about a bat's perceptions is that, since bats navigate in their dark caves by a sonar-like system, they will perceive a range of qualities which we do not perceive. However, even where human and bat senses more or less overlap, the way of life of the two species is so utterly different that the quality-organization or quality-space of a bat's perceptual field is sure to be quite different from that of a human being. And place in a different quality-space constitutes a difference in quality. (If orange was not between red and yellow, then it would not be the quality it is.)

Presumably the bat's perceptions involve secondary qualities. (We could imagine that through some science-fiction hooking-up of the central nervous system of bat and human being it becomes possible for us to perceive using the bat's sense-organs and perceptual processing. We might then perceive the qualities perceived by the bats.) If an account of secondary qualities in terms of primary qualities is correct, the bat in all cases will be perceiving purely physical features of the environment. But, of course, the features would not be perceived *as* the physical features that they are.

In order to appreciate the vast difference between a bat's 'point of view' and our own it may be useful to consider the following case where the difference is not great at all. A puzzle-picture can contain many lines, some of which make up a face. The problem is to see this face. In an ordinary situation we will not know which are the lines that make up the face. We can imagine, however, that we are actually shown which the face-lines are, that they represent a face, and that we are able to keep the lines in mind and indicate them at any time, perhaps by tracing them out. One might even imagine being shown the drawing of the face out of its confusing puzzle-context, and so coming to know just what sort of face it is. Yet, with all this, it is possible that we should still not see the face in the picture *as a face*.

Now suppose that we do come to see the picture as a face. Suddenly the lines seem transformed.

The point to notice in the illustration is how small a change has occurred in our perceptual organization. At first we cannot, and then we can, take the face-lines in the picture as a natural unit. We come to treat a thing as a perceptual unit which previously we did not treat as a unit. It is a classificatory acheivement which in this case does not even yield us any new knowledge. (We knew what lines made up the face, and even knew what sort of face it was.)

In this illustration, I do not think that anything turns upon the fact that the lines are not in fact a face, but only represent a face. The lines might do no more than pick out a certain complex but organized shape actually present on the page. But still there is a contrast between knowing that the lines form a certain shape, and actually seeing that shape there on the page in a unified way.

Now if so *small* a change in perceptual organization is so striking, then who can doubt that if we could have experience of the perceptual organization of other species, or even of some persons of our own species, then we would, as it were, be taken into a new world? We would have a completely new point of view.

Yet there is nothing in the puzzle-picture case to disturb the causal theory of the mind. Equally, I suggest, nothing in the fact of more radical differences in perceptual organization need disturb the causal theory.

The inverted spectrum

The apparent possibility of an inverted spectrum, that what looks red to you looks green to me and vice versa, may seem to pose greater difficulties for the causal theory. Suppose that, perhaps under the influence of the Argument from Illusion, one took it that looking red and looking green were a matter of the sensible qualities of redness and greenness qualifying something mental. It then seems that two persons could have their colour-perceptions reversed relative to each other, yet this to be a reversal which could not be detected in behaviour.

A purely causal theory of the mind, however, which denies that there are any introspectively experienced qualities, cannot give this account of colour-reversal. What account can it give? If it has to deny that the reversal is a possibility, that is a difficulty for the causal theory. For, *pace* Malcolm among others (pp. 60–1), the case seems to be intelligible.

The intelligibility of the case is supported by the following consideration. It is surely possible that an individual should suffer colour-reversal. Suddenly red things look green to him and green things look red to him. Such a switch would be immediately evident to him, and would have detectable results on his behaviour. But if such a switch is possible, then it is hard to deny that different persons' colour-perceptions may be reversed relative to each other.

It is possible to doubt whether the two-person case would be a case where the difference was behaviourly undetectable. For behavioural undetectability it would be necessary that the quality-space in which the transposed colours were set had a certain symmetry. Each relationship that red had to other colours would have to correspond in its formal properties to a relationship that green had to other colours, and vice versa. Otherwise, sorting and matching tests could reveal that the qualities involved for the two perceivers were not the same. Now, as a matter of fact, it is not clear whether the structure of colour-space does have the requisite symmetry. The situation becomes even more complicated if cross-sensory comparisons (for instance, between the 'warm' and the 'cool' colours) have to be allowed for.

However, it may be advisable to set this doubt by. For whatever be the case about colours, it seems clear that there might be beings with quality-spaces with the requisite symmetry. If so, it seems that the perceptions of one of these beings might be inverted with respect to the other, but with the inversion behaviourly undetectable. So let us assume that colours have the right symmetry.

At this point it is desirable to issue a *caveat*. While it may be possible to make intelligible the supposition that A's colour-perceptions are reversed relative to B's colour-perceptions, despite undetectability in behaviour, it is not possible, at least given the causal theory, to say that one of the two is 'seeing the world rightly'. Suppose the following very primitive theory of perception. Veridical perception of a certain quality consists of coming to have something with the same quality. Given this theory, if A's spectrum is inverted relative to B's spectrum, then at least one of the two persons must be misperceiving the world. For the causal theory, however, both A and B will be perceiving the colours correctly. Any talk about who has the 'right' colour-perceptions could only be a matter of which of them was physiologically normal. And if the population were equally divided between A-type and B-type perceptions, then there could not even be appeal to physiological normality. Inversion can only be a *relative* matter.

But this leads on to a difficulty. What A has when he looks at a red surface under standard conditions is a perception which the causal theory can only call the surface looking red to A. Similarly when A looks at a green surface. The same goes for B. But, if so, what sense can we make of A and B having inverted spectra relative to each other?

I think that we can nevertheless make what one might call 'sufficient' sense of the inverted spectrum. What we want is the logical possibility that A's looking red is B's looking green, and that A's looking green is B's looking red. And that possibility, it seems, we can describe.

Let us go back to the case where a person suffers colour-reversal, a reversal he is aware of. How should we describe it in causal terms? Something like this presumably. Those objects which previously stimulated red-selectors in the subject's mind are now stimulating green-selectors, and vice versa. If we are thinking in materialist terms, there has been a switch in the relevant brain-mechanism. Given no compensating switch in the introspective apparatus, then the subject would be aware of what had happened. But in time he might get used to it, and might eventually say that red things once again 'look red' to him.

Now suppose that this switching possibility exists in everybody's brain. The person who switched from state A to state B did so in mid-life. Normal persons, however, spend their whole life either in state A or in state B. Now consider the situation of persons in state A or B with respect to persons in state B or A. Can we not say that there is *a* clear sense in which, relative to each other, they have inverted colour-perceptions? If so, we have made sufficient sense of the logical possibility of inversion while accepting the causal theory.

Albert is in state A, Bertrand in state B. The following counterfactual holds about Bertrand. If he were to switch to state A, then red things would stimulate what had been his green-sensitive elements. It is just these sorts of elements, though, physiologically speaking, which are red-sensitive elements in Albert. If these things were all true, and it is logically possible that they might all be true, would they not constitute *a* sense in which Albert and Bertrand had reversed colour-perceptions?

So I think that the causal theory can accommodate the logical possibility of inverted spectrum perception without any 'input-output' differences correlated with the inversion. Putting the matter at its lowest, *enough* sense can be made of such inversion to explain why it seems a pre-theoretical logical possibility.

At the end of his essay Malcolm suggests that the truth about the mental lies plainly before us. All we have to do is avoid being bewitched by theory. My own view is less optimistic than this. I do not see that we can avoid theory when we think about the mental. But some theories are better than others, and observation, discussion and argument may, in the long run, favour one theory rather than another.

Norman Malcolm's Reply

Bodily sensations. In his essay Armstrong says that 'there is some essential or definitional link . . . between pain and the natural behavioural expression of pain' (p. 107), that pain is 'a species of bodily sensation which characteristically evokes a very direct desire that that sensation should cease' (p. 113); that 'by definition, severe pain is a sensation which arouses strong adverse reaction' (p. 133); that 'It is of the essence of pain that it is a bodily sensation, and that, in normal circumstances, this bodily sensation creates in us a most peremptory desire to be rid of it' (p. 143).

In those passages Armstrong seems to be holding that there is a conceptual connection between pain and the behaviour of expressing pain. But Armstrong also seems to hold the view that the relation between pain and its behavioural expression is only contingent. In my essay I quote Armstrong's remarks (in his *A Materialist Theory of the Mind*) about a phenomenon that sometimes occurs after prefrontal lobotomy is performed on patients to relieve severe and intractable pain. What Armstrong says there is that sometimes patients who have had this operation will 'say that the pain is still there, but it does not worry them anymore. It seems as if they are saying that they are having a pain which gives them no pain.' My comment in my essay is that 'If a patient said that the pain was as *intense as ever* but that he didn't *mind* it anymore, then we should not understand what he is saying' (p. 12). Armstrong, in his essay, finds this 'incomprehension puzzling'.

> For surely physical pain involves at least two elements: (1) bodily sensation of a certain sort or sorts; (2) an unfavourable reaction to the having of these sensations. (Contemporary psychology distinguishes in a fairly routine way between the bodily sensation of pain, on the one hand, and the distress or suffering it causes, on the other.) But if pain involves these two elements, why should not the two be dissociated in some special circumstances? (p. 114)

Of course, if a lobotomized patient reports that he still has some pain but that it isn't severe enough to distress him, that would be an intelligible report. Also, a person in severe pain might have a *motive* for suppressing any expression of pain, e.g. he does not want to appear unmanly. Someone might even *rejoice* in at last having pain in a limb in which he had previously lost all feeling: in this sense he would have a 'favourable reaction' to the pain in that limb, not an 'unfavourable' one. Or if a man undergoes a necessarily painful treatment, which he realizes is essential for saving his life, he may have no 'desire that the pain should cease' before the treatment has been completed. So, there can be cases in which pain is welcomed; cases in which there is no desire for it to cease; cases in which pain-behaviour is suppressed.

But, to speak tautologically, pain is painful. What we cannot coherently think is that physical pain involves 'two elements', one a bodily sensation (pain), the other the painfulness of the sensation, *and* that these two could be 'dissociated in some special circumstances'. It is a contradiction to suppose that a person might *have* severe pain but not *suffer* any pain—despite what 'contemporary psychology' might say to the contrary.

A person suffering pain has a tendency to express this suffering in words or other behaviour. That is a conceptual truth, not a contingent one. Someone can be in pain without betraying it in any way—but if mankind had never exhibited pain-behaviour there would be no *concept* of pain.

The causal theory creates confusion by regarding pain as a sensation that typically 'causes' pain-behaviour, where 'the causal relation' is viewed as a contingent relation. But (to repeat) the connection between pain and the natural expression of pain is conceptual, not contingent. There are a few human beings who have *no* tendency towards pain-behaviour when their bodies are burned or their limbs broken. What is said of them, and has to be said, is that they do not have sensations of pain. It makes no sense to say that perhaps they have the *same* sensations that normal people have but with them those same sensations do not 'cause' pain-behaviour.

According to Armstrong the causal theory holds that 'Mental phenomena have natures of their own, and *considered as things having that nature*, there is no necessity that they should have any particular causal powers' (p. 147). Sensations, thoughts, emotions, intentions, are said by Armstrong to be 'mental phenomena'. What can it mean to say that they 'have natures of their own'? A bodily sensation of pain is described in terms of *location* ('in the hand'), *quality* ('throbbing'), *degree* ('slight', 'intense'), *duration*, and *frequency*. But this familiar sort of description is *not* what Armstrong would mean by a description of a bodily sensation 'in its own nature'. I will return to this point shortly.

It is striking that Armstrong has, except for his materialism, the same view about pain as does Descartes. Armstrong says:

> We say that we have a pain in the hand. The *sensation* of pain can hardly be in the hand, for sensations are in minds and the hand is not part of the mind. A brain kept functioning without its body might suffer pain if stimulated in certain ways. (p. 182)

This brief passage is a jumble of philosophical errors. It implies that we are always *mistaken* when, in responding to the question 'Where do you have the sensation of pain?' we say that it is in the hand, or in the leg or shoulder. Presumably what one should say is, 'It is in my mind'. Of course this is all wrong. In learning the language of sensation we learn to give the bodily location of a sensation. If a child who has not yet learned to speak is in pain, we can determine from his natural pain-behaviour (flinching, limping, caressing, pointing, crying out when a certain place is pressed) the location of the pain. This behaviour is directed towards a certain part of his body, and it *defines* for us the location of his pain.

Not all pain has a bodily location: there is the pain of grief, of anguish, of humiliation. The pain of humiliation can be called 'mental pain', but not the pain of a twisted ankle: this pain is in the ankle, not 'in the mind'.

It is nonsense to say that a brain without its body 'might suffer pain' if stimulated in certain ways. It is *people* (and animals) who suffer pain, not brains. And, as far as I know, people do not have sensations of pain in their brains: I have been told that the brain is actually insensitive, unlike most other parts of the body.

Intentions. Armstrong sees that there is an essential or conceptual link between purposes or intentions, and the actions of fulfilling them. He says: 'If it is somebody's objective to open the door, then, by and large and in the absence of obstruction, door-opening behaviour will follow' (p. 107). Also: 'If it were not true that those who had simple and straightforward purposes are generally able at least to do things conducive to realizing them, then those purposes could not be said to be purposes, or not said to be the purposes which they are' (p. 142). In general, he says, 'there is a conceptual connection, a subtle form of necessary connection, holding between mental phenomena and behaviour' (p. 147).

This account of the relation of intentions to actions is largely correct. But, as in the case of bodily sensations, Armstrong has another view that seems to contradict this one. Let us return to his notion that 'mental phenomena' have 'natures of their own', and consider this idea in its application to intentions. I ask you what you intend to do this afternoon, and you answer 'To mow the lawn'. This is a familiar and intelligible description of your intention, and I know what to expect. But, according to Armstrong, you have not described your intention 'in its own nature'. What is your intention 'in its own nature'? According to Armstrong's materialist hypothesis, it is a certain brain-state. Neither I, nor most people, know how to describe brain-states; so we are unable to describe our intentions 'in their own nature'. It is supposed that someday neurologists may be able to do this.

Armstrong is engaged in a gallant effort to reconcile two positions that appear to be irreconciliable. The first position is his correct observation that there is a conceptual, non-contingent, connection between an intention and the actions of carrying out the intention. The second position is his causal theory of intention, according to which an intention 'causes' the behaviour of fulfilling the intention, in a sense of 'cause' in which a cause and its effect are 'distinct existences', i.e. are logically independent of one another: this causal relation is purely contingent. When these two positions are conjoined we get the result that your intention to mow the lawn is connected *both* contingently and non-contingently with your action of mowing the lawn. This seems contradictory. Armstrong is aware of this appearance of contradiction and seeks to remove it by invoking what I call his double description theory. According to it a purpose or intention can be described in two different ways: first, in the familiar way as, e.g. 'the intention to mow the lawn'; second, 'in its own nature' as, e.g. a certain brain-state. Armstrong says that

> in the case of purposes we can conceive of them as 'distinct existences' from the thing which they bring about. The purpose, characterized as a purpose, is described in terms of what it tends to bring about. But it is possible to characterize it independently, for instance, if one accepts materialism, as a brain-state or process. It is then clear that it *is* a distinct existence from the coming to be of the state of affairs purposed. (p. 168)

I drew from this view a consequence that seems absurd, namely, that a particular intention (e.g. to mow the lawn) *might* have been a different intention (e.g. to work in the office). That this is a consequence is readily seen. Armstrong holds both to materialism and to the causal theory. According to his materialism your intention to mow the lawn (call it 'M') just *is* a certain brain-state (call it 'B'): M and B are identical. According to Armstrong's causal theory an intention is *defined* in terms of the behaviour it tends to bring about: thus your intention, M, *is* the intention to mow the lawn. But according to the causal theory causation is a contingent relation between the cause and its effect. In a different causal set-up your brain-state, B, would have caused the behaviour of working in the office, and therefore would have *been* the intention to work in the office (call it 'W'). As things are, M and B are identical. But as things might have been, with a possible but different causal set-up, B and W would have been identical. Things equal to the same thing are equal to each other. Therefore M and W might have been identical: your intention to mow the lawn might have been your intention to work in the office.

This does *not* mean that instead of having your present intention you might have had a different intention. It means that that very intention, to mow the lawn, could have been the intention to work in the office. This is a topsy-turvy consequence for the concept of intention. In the actual employment of this concept in thought and language, intentions are identified solely by *what is intended*—that is, by the *object* of intention. Intentions with different objects are ipso facto different intentions, and could not have been the same intention. My hand can hold different things at different times but remain the same hand. But an intention does not have 'a nature of its own' which permits it to be the *same* intention with different objects.

In his essay Armstrong responds to this criticism by acknowledging that it is a consequence of his theory that a certain intention might have been a different intention: e.g. the intention to paint the bathroom might have been the intention to write a poem. He says that this consequence can be 'calmly accepted' (p. 148). He thinks that the difficulty in conceiving this is due to the fact that an intention belongs to a complex causal network. He says:

> I cannot have the intention to paint the bathroom, or the intention to write a poem, unless I have at least the concepts of paint, bathroom, poem, *etc.* with all that these sophisticated concepts involve. As a result, to suppose that the cause which is in fact my intention to paint the bathroom logically could be an intention to write a poem, is to suppose a complex and messy transformation of indefinitely many elements from playing one sort of causal role to playing another. . . . No wonder, then, that the logical possibility of the transposition seems so problematic. (p. 148)

This explanation of the difficulty is not adequate. According to Armstrong's materialism every particular 'mental state' is a particular brain-state, and the causal links between these various brain-states are contingent. To have the concepts of bathroom, paint, etc., would be to have certain brain-states; the intention to paint the bathroom would be another brain-state; the causal connections between these various brain-states would be contingent. When Armstrong says that he cannot have the intention to paint the bathroom unless he has the concepts of paint and bathroom he has to mean, on his view, that this is causally impossible, not logically impossible.

Here I will draw a point from a recent article by Elizabeth Anscombe. She is criticizing the notion that a particular brain-state might be a *sufficient condition* of a certain belief. (On Armstrong's view, not only intentions, but also beliefs, thoughts, decisions, desires, etc., are brain-states. If a certain belief *were* a particular brain-state, then the presence of that brain-state would be a sufficient condition for the presence of that belief). Anscombe remarks that 'there can be no such kind of brain-state as *the* kind of brain-state corresponding to such-and-such a belief in the sense of being a sufficient condition of it'.[1] For example, a certain brain-state could not be a sufficient condition for a belief about banks on the part of a person who had never heard of a bank. Anscombe imagines the following reply: 'The brains of such people never do get into any of these states. The causal conditions for getting into them exist in nature only where there are banks, etc.' Anscombe makes this rejoinder:

> But let us suppose a way of producing one of these states artificially, i.e. outside the circumstances in which the causal conditions occur 'naturally'. And now, consider the inference that if such a state has been so produced the subject is then in a state of belief that, say 'such-and-such a bank in—cester is open at 5.00 P.M. on Thursdays', though neither—cester nor banks nor clocks nor days of the week ever came into his life before, nor did he ever hear of them. The absurdity of the inference brings it out that . . . the brain-state is not a sufficient condition for the belief.[2]

This point applies equally to the example of the intention to paint the bathroom. The impossibility of a person's having this intention who does not have the concepts of paint and bathroom is *logical*, not causal.

Theoretical entities. Armstrong holds the curious view that mental states are 'theoretical entities postulated within ordinary language and thought' (p. 143). Since for him, sensations, thoughts, intentions, decisions, are 'mental states', then all of these are theoretical entities. More accurately, they are partly observational and partly theoretical entities:

> They are semi-observational, semi-theoretical entities. They are observed. But what they are observed *as* is simply as things

[1] G. E. M. Anscombe, 'The Causation of Action', in *Knowledge and Mind*, edited by C. Ginet and S. Shoemaker, Oxford University Press, 1983, p. 182.

[2] Anscombe, 'Causation of Action', p. 183.

> which play a causal role. We are often directly aware of our own purposes, for instance. But all that we are aware of them *as* is as states within us impelling us in certain directions, the direction being the state of affairs purposed. (pp. 145–6)

This is an extraordinary way to speak about an intention. If I intend to mow the lawn, am I aware of this intention 'as a state within me impelling me towards mowing the lawn'? If my neighbour asked me what I intend to do this afternoon, would I answer: 'There is a state within me impelling me to mow the lawn'? Such a reply would seem to be a description of a neurotic impulse, instead of an ordinary announcement of my intention.

Is there such a thing as *observing* one's own intention? What is there to observe? I can tell you what I intend to do: but my announcement of my intention is not derived from any observation. I will probably have a *reason* for my intention, and the reason might refer to an observation, e.g. that the grass is getting too high: this would not be an observation of something *within me*, but of something out there.

Armstrong's notion that an intention is a theoretical entity assumes that I *observe* my intention, but that I observe it only *as* something that causes, impels, drives me to do what I do. My observation does not disclose the nature of this something; it is *something I know not what*: the real nature of this something is a topic for scientific theory and research.

There are several errors in this view. The first error is to think that my intention to do X causes or impels me to do X. The second error is to think that in the normal case one observes or perceives one's own intention. The truth is that sometimes I can perceive another person's intention; but my normal ability to say straight-off what I intend to do is not due to my observing or perceiving my intention. The third error is to think that an intention has 'a nature of its own' which might be studied by science.

Armstrong's conception that an intention 'in its own nature' is a brain-state and that this brain-state causes the behaviour of fulfilling the intention, seems to lead to the following paradoxical consequence: the average person knows nothing about his brain-states or their causal role in relation to his behaviour, and also realizes that he is ignorant of these matters. Suppose that such a person seriously believed that an intention is a brain-state. If asked what he intends to do this afternoon, would it not be reasonable for him to reply, 'How should I know? Let us wait to see what I will do.' Later he mows the lawn, and he *infers* from this that his previous intention was to mow the lawn! This would be a travesty of the employment of the concept of intention. If it ever came about that Armstrong's view seriously influenced our thought and language, the consequence would be a *dismantling* of our present concept of intention—which means that Armstrong's view cannot be a correct account of this concept *as it is*.

Genuine duration. I called attention to a particular concept of duration that Wittgenstein terms 'genuine duration'. An example of something that has genuine duration is the rolling of a ball across the floor, or the rising of a liquid in a tube. One could be asked to keep the ball under observation and to keep reporting whether it was still rolling; when it had stopped; when it started rolling again, etc. The same for the rising of a liquid in a tube. The concept of genuine duration is tied up with the idea of keeping something under constant observation, of focusing one's attention on it and being on the alert to report or signal any change in its quality or state. Brain-states have genuine duration, since they could be monitored continuously by instruments which would record changes in chemical balance or quantity of electric charge in specific areas of the brain, or changes in the firing pattern of a cluster of neurons, etc.

But an intention does not have genuine duration. You may have had a certain intention for a month or a year; but an intention is not a sensation, image, or feeling: it has no experiential content, nothing that comes and goes, emerges and fades—such that you could report, 'Now it's there; it's still there; now it's gone'. In an intention there is nothing to observe or to keep under observation. I conclude in my essay that an intention cannot be a brain-state, since a brain-state has genuine duration but an intention does not.

Armstrong argues in his essay that if an intention does not *lapse* then it continues in existence, and therefore it has genuine duration (p. 159). This is a *non sequitur*. To say that someone's intention, e.g. to study law, has 'lapsed', would presumably mean either that the person has given up the intention, or has forgotten it, or perhaps has postponed it. If his intention has not 'lapsed' then he still has it. But it doesn't follow from this that his intention has genuine duration. Genuine duration can be attributed only to something that can be an object of attention, or of constant observation, or of continuous monitoring. An intention is *completely* described by saying *what is intended*, e.g. to study law. An intention has no other content. There is nothing to observe, focus one's attention on, to monitor. Therefore, genuine duration cannot be attributed to an intention.

Qualities of experience. In section 4 of his essay Armstrong considers the view of some philosophers that 'mental phenomena involve a qualitative component not to be captured by any analysis in terms of mere causal role' (p. 169). I treated this topic in section 2 of my essay, but Armstrong seems to have misunderstood my position. He says:

> Malcolm also criticizes the notion of introspective inner qualities. But I am in fact much more sympathetic to the idea that there are such qualities, than I am to Malcolm's criticism. Malcolm is hostile to the whole idea of introspection and qualities perceived only by the introspector. I have no such hostility. All I shall be arguing is that *in fact* there are no such qualities. (p. 169)

But I am not hostile to the idea of there being qualities of experience. I said, for example, that there are genuine qualitative characterizations of bodily pain. A pain can be described as 'piercing', or 'dull', or 'gnawing', or 'throbbing': these would be descriptions of the *quality* of a pain. If a lorry driver had the experience of being marooned in his lorry for twenty-four hours in a blizzard, he might later tell me that he was in an agony of fear that he would freeze to death before he was rescued. He would be telling me something of the subjective character or content of his experience of being marooned—of 'what it was like for him'.

What I am doing in my section 2 is criticizing the way in which some philosophers *use* such phrases as 'the subjective character of experience', or 'the qualitative character of a mental state'. They speak of 'the qualitative character' of a headache, or of fear, or of seeing—but they do not go on to specify what that 'qualitative character' is. They then claim that

that 'qualitative character' cannot be captured by any analysis in terms of causal role. One point I make is that there can be no responsible judgement that their claim is either true or false, as long as they do not *specify* what these 'qualitative' or 'subjective' characteristics of experiences or mental states *are*. I go on to propose the conjecture that the explanation of why no such specification is offered is that these philosophers have assumed that they can specify the qualitative character of seeing or fear or pain, by inner ostensive definition, by inward pointing —whereas in fact *inward* pointing cannot fix the reference of any term.

In section 4 of his essay Armstrong puts forward a number of interesting proposals, one of them being that perceived secondary qualities really are primary qualities. Since these proposals have no bearing on my essay, I hope I will be pardoned for not attempting to discuss them here.

Causality. In his essay Armstrong repeats his view that a cause and its effect are always 'distinct existences'. I argue in my essay that the terms 'cause' and 'effect' are used in ordinary language and thought to refer to a variety of different relationships. One example I give is of a chess player making an unorthodox move in order to confuse his opponent. The ploy is successful: his opponent does become confused. Compare this with the case in which the opponent is secretly given a drug, the effect of which is that he becomes confused. Compare the two sentences, 'The strange move caused him to be confused', 'The drug caused him to be confused'. The same word, 'caused', occurs naturally in both sentences: but we are being presented with *different concepts* of *cause*. The difference becomes apparent when we consider that in the first case the confusion has an *object*, namely, the other player's strange move. The latter is *both* the cause and the object of the confusion. It is a causal relationship in which the cause and the effect are internally connected: the description of the effect involves the description of its cause. In the second case the drug is only the cause and not the object of the player's confusion; his confusion does not have an object: he is not confused or bewildered *at* anything. These are plainly different concepts of causal relationship, and cannot be dismissed as mere 'epistemological differences'.

Armstrong acknowledges that the player's being disconcerted, confused, bewildered, 'is a mental state with an intentional object', the intentional object being the other player's strange move. Apparently Armstrong is conceding that under the description, 'The strange move caused him to be disconcerted', the cause and its effect are not 'distinct existences', for he supposes that this example should be treated in the same way that he treats intentions and purposes. This manœuvre consists in resorting to his materialist thesis that mental states are brain-states. The second player's 'belief' that the first player made the chess move he did make, is a certain state of his brain; his 'belief' that the move is a strange one, is another state of his brain; these two brain-states jointly cause him to be disconcerted, which is a third state of his brain. All three of these brain-states are 'distinct existences', and so the causal relation between them is purely contingent (p. 168).

My response to this is to apply Anscombe's point on which I drew previously in this rejoinder. It is logically possible that these three brain-states could be produced artificially, i.e. outside the circumstances in which the causal conditions occur 'naturally'. Let us suppose they are produced in the brain of a person who has never seen, played, or even heard of chess. Can we infer that this person (now sitting in front of a chess board) believes that the person opposite him made *such-and-such a move in chess*; and believes that this move is *highly unorthodox*; and is disconcerted by *the strange character of this move*? Of course not! The absurdity of the inference shows that no brain-states can be sufficient conditions for such beliefs, or for such a state as that of being disconcerted by a certain move in chess; and *a fortiori* it shows that such beliefs, etc., cannot be *identical* with any brain-states. Since Armstrong's materialist thesis is false on logical grounds, it cannot serve to bolster up his causal theory of mind.

D. M. Armstrong's Reply

I will go through Norman Malcolm's rejoinder, trying to answer him point by point.

Bodily sensations. After quoting four passages from my essay, Malcolm goes on to say that I seem to hold that there is a conceptual connection between pain and the behaviour of expressing pain. Actually, I say this in only one of the quotations given. The other three quotations assert the existence of a conceptual connection between pain and a *desire to be rid of it* or *an adverse reaction to it*. It is true that I think that there is a (subtle and complex) conceptual connection between desires (and adverse reactions) and behaviour, and so, by transitivity, a connection between pain and behaviour. Nevertheless, I think my qualification is worth making. For in contemporary analytical philosophy of mind there is too great a haste to establish conceptual connections between the mental and behaviour. The internal conceptual connections between the various mental phenomena are then in danger of being overlooked, to the ultimate detriment of the enterprise of showing the links between these phenomena and behaviour (see pp. 153–7).

Malcolm goes on to say that I seem to hold that the relation between pain and its behavioural expression is 'only contingent'. My reaction to this comment is again ambivalent. I think that the question whether a relation between items is contingent or necessary is relative to the descriptions or concepts under which the items are brought. Under the description 'pain' and under ordinary descriptions of behaviour, the relation between pain and *some features* of its behavioural expression seems not to be contingent. Nevertheless, pain and its behavioural expression are 'distinct existences' in Hume's sense. (If this distinctness of existence is doubted, then the case of the curare-induced paralysis should put it beyond doubt. The paralyzed person can be in great pain, yet behavioural expression be completely denied to him. See section 1 of my essay.)

A useful, if considerably over-simple, analogy is the relation of father to child. Given that we employ the descriptions 'father' and 'child', then it is not contingent that a father is father of some child. (Though contingent that the father is father of *this* child.) Nevertheless, the father and the child are 'distinct existences' and, it may be noted, the father is (part) cause of the child. The father can of course be described in ways which make it contingent that the being so described is part cause of the existence of any child, for instance as a man. My view is that pain stands to its behavioural expression (where the latter occurs) something as a father stands to his child. In its own nature, I hypothesize, the pain is a brain-process, and, described as a brain-process, the connection between pain and its behavioural expression ordinarily described is purely contingent.

Malcolm goes on to discuss the case of lobotomized patients who report that they still have pain, but say that it does not worry them any more. His position seems to be that such persons are not in pain. He says that we cannot coherently think that physical pain involves two elements, a bodily sensation and the painfulness of the sensation, *and* that the two elements could be dissociated in some special circumstances. I suppose his point is that one cannot *define* an A as *an F which is a G*, yet go on to claim that, in some circumstances, we can find As which are Fs but are not Gs.

The point is formally correct, of course. But what happens if certain sorts of sensation evoke certain unfavourable reactions, are indeed named 'pains' from these reactions (from the Latin *poena*, penalty, punishment, (later) pain, grief—OED), but then, unexpectedly, are found sometimes to occur without the reactions? How is the person who has such a sensation, and is aware of having it, to describe it? There is a problem here. The lobotomized patients solve the problem by calling these sensations 'pains'. They are a bit like people who call chess without the queen 'chess'. We could take the pedantic point that the formal laws of chess demand that each side have a queen. Yet chess without these queens has such a strong family resemblance to chess that it is not unreasonable, even if not compulsory, to call it 'chess' also.

I am assuming that the lobotomized patients are able to recognize their sensations as being the same sort of sensation as those which, on previous occasions, evoked the unfavourable reaction involved in normal cases of pain. But that assumption does not seem unreasonable, although I suppose that Malcolm would contest it.

Malcolm may be right in saying, as he does, that it is a contradiction to suppose that a person might have *severe* pain, yet not *suffer* any pain. The word 'severe' here seems to import suffering. But it seems that a lobotomized patient might be able to recognize that he was having a sensation of a sort which, before the lobotomy, would have made him suffer severely.

Malcolm goes on to consider the case of those unfortunate human beings who have no tendency towards pain-behaviour when their bodies are burned or their limbs broken. It is nonsensical, he claims, to say that they might have the same sensations that normal people have, but yet that these sensations should not 'cause' behaviour (his inverted commas). For myself, I should be a bit surprised if they did have such sensations. But I cannot understand why he is so dogmatic. Neurophysiological and other research might make it plausible that they were more like the lobotomized persons than one would be inclined to suspect.

I am glad that Malcolm has noticed the similarity of my view of pain to Descartes's view. I discovered the resemblance myself, in Descartes's *Passions of the Soul*, after working out my own position. Perhaps the similarity is not a complete accident. Descartes holds that the mind is a spiritual substance. I hold that it is a material substance. We have common opponents: those, such as Malcolm, who hold that it is not a substance at all.

Malcolm goes on to say that, according to me, if somebody says that they have a pain in their shoulder, then they are *mistaken*. For, according to me, pains are sensations, sensations are mental, what is mental is in the head, and not in the shoulder.

I reply that this is an extremely unsympathetic way to take my remarks. Naturally, there is a clear sense in which I hold that such remarks as 'His pain is in his shoulder' are often true. But what are the truth-conditions for such statements? I think that their truth-conditions are utterly different from, say, 'The bullet is in his shoulder' or 'Physiological process P is going on in his shoulder'. (So, I am sure, does Malcolm!) I hold that to say that he has a pain in his shoulder is to say that it feels to him that a disturbance of a certain idiosyncratic sort is going on in his shoulder. The 'feels' here is the 'feels' of bodily perception, and can be a non-veridical perception. The perception characteristically evokes the desire that the perception should cease. If this analysis is correct, then the 'place of the pain' is a merely intentional location, although, if the perception is veridical, there *is* a special sort of bodily disturbance at that place. To say that the pain is in the shoulder is then compatible with, indeed entails, that the sensation, and mental reaction which it evokes, are both 'in the mind'. If materialism is correct, then they are in the brain. What is more, saying that the pain is in the mind seems, in the context of philosophical discussion, rather more illuminating than saying that the pain is in the shoulder. *Pace* Malcolm, and agreeing with Descartes, 'bodily' pain is like the mental pain of grief, anguish, and humiliation.

Malcolm next remarks that it is nonsense to say that a brain without its body might suffer pain. People and animals suffer pain, not brains. The brain itself is insensitive.

With regard to the insensitivity of the brain, we must distinguish between casual stimulation and the stimulating of different pain-pathways, or pain-centres, in the brain. The latter stimulation can be painful enough, although the pain is not felt to be in the brain.

Malcolm's position here reminds me of the Cartesians and their doctrine that animals are automata, and do not have sensations, much less thoughts. I understand that this led some of them to be extremely brutal to animals. The dog might whine and yelp dismally, but it was not really feeling anything. So you could kick it or cook it alive in an oven in good conscience.

I understand that the disembodied brains of monkeys have already been kept biologically alive for short periods. They exhibit EEG rhythms of a sort correlated with certain mental conditions in men and animals. What would Malcolm think of the stimulation of the pain-receptors, or the pain centres, of such disembodied brains? I hope that his faith in his view of mind would not be so strong as the Cartesian faith in the doctrine of the brute-machine.

Intentions. In the section of Malcolm's reply dealing with intentions we seem to go round much the same mulberry bush that we have just gone round in the case of pain. Taking yet a further turn around the bush, I would remark that my 'double description theory', as Malcolm dubs it, is not a particularly paradoxical affair in other contexts. Described as a father, it is a necessary truth that the being so described engenders a child. Described as a man, it is not necessary. Described as a set of genes, it is a necessary truth that this object is causally apt for the production of certain hereditary characteristics. Described as a DNA molecule, it is not necessary. So why not a double description in the mental sphere, with the mental conceptually connected to its behavioural expression, and yet with the mental totally distinct from its behavioural expression?

However, Malcolm thinks that the theory is paradoxical in the mental context. In particular, he argues, I am forced to maintain that, e.g. an intention to mow the lawn, that very intention, might have been an intention to work in the office. This, he says, is absurd.

But why is it absurd? Malcolm's answer is that, as we actually employ the concept of intention in thought and language, intentions are identified *solely* by what is intended, by the object of the intention.

I cannot see that I can do much here except give Malcolm the lie direct. I agree that intentions with different objects are *ipso facto* different intentions. But it is clear that there are many descriptions of an intention available which cut across the question of the object of the intention. (For instance, a certain intention can be truly described as the thing which we were talking about at 2.10 p.m. yesterday.) I say further, against Malcolm, that there is *room* (logical space) for a description of intentions as material states of the brain. That description is not logically barred. And if not barred it seems to be a scientifically plausible identification.

Malcolm thinks that this is absurd, though. How could we begin to settle the issue between us? I suspect that the only approach which has any promise at all would be to consider our respective views in a larger context. We would need to step back and consider philosophical methodology, scientific plausibilities, and so on.

If I am correct about intentions, then, given that intention J is a certain brain-state, it is logically possible that that very brain-state might have been intention K, an intention with a different object. In my original essay I admitted that in the case of fairly complex and sophisticated intentions such a logical possibility can seem problematic, because of the complex and messy transformations that it would involve. The brain-states which are in fact the very complex system of concepts involved in the intention to paint the bathroom would have to be conceived of as the system of concepts involved in the intention to write a poem. It does not sound particularly plausible.

Malcolm says that, nevertheless, on my view the transformation would still have to be logically possible. This, however, I grant. All I was trying to do was to show why it might not *seem* possible to some philosophers. As the history of philosophy has shown again and again, and as I am sure Malcolm would admit, we can have *illusions* of logical impossibility and possibility. Furthermore, it does not seem particularly *im*plausible to say that in this case there is no more than the illusion of logical impossibility. For we can see why the illusion should arise.

Malcolm goes on to mention, and endorse, a difficulty recently raised by Elizabeth Anscombe:

> . . . there can be no such kind of brain-state as *the* kind of brain-state corresponding to such-and-such a belief in the sense of being a sufficient condition for it.

The point is illustrated by saying that a certain kind of brain-state could not be a sufficient condition for a belief about banks.

The point made by Miss Anscombe is not one which I wish to contest. I made this clear, I think, in pages 160–3 of my essay, where I rejected the necessity of a *type-type* identification of mental states with states of the brain. A belief, according to me, is a structured state. This state is constituted the particular sort of belief that it is (belief that *p*) in virtue of the complex causal role which it plays, a role involving its causes, its causal relations to other mental states (particularly important in the case of beliefs) and its causal relations to behaviour. 'Causal relations' here may involve a wide variety of relationships, variations on the theme of (ordinary, efficient) causality. But there is no reason why there should be a certain sort of brain-state which is type-identical with the beliefs which involve banks. Different sorts of structure may play the same causal role in different minds.

It is clear, of course, that if the belief is to be a belief about banks, then, analytically, and therefore necessarily, it must involve the concept of a bank. (So I do not reject Malcolm's point that the impossibility of somebody having the intention to paint the bathroom, but lacking the concepts of paint and bathroom, is logical, not causal.) Hence, if a token of a belief about banks is to be a brain-state, it must link up in a certain way with a token of a brain-state which can be identified with a concept of a bank. (For details, see my *Belief, Truth and Knowledge*, chapter 5, section I, where, following but adapting Peter Geach, I draw a distinction between concepts and Ideas.) Furthermore, the concept of a bank cannot be an isolated object. Of logical necessity, it must be linked in various ways to a great multitude of other concepts. If all these further concepts are to be brain-states, then the brain-state which is the concept of a bank must be suitably linked to further brain-states, identified as the concepts which 'surround' the concept of a bank.

But at no point, it seems, do we require the notion of a *kind* of brain-state which, whenever it occurs, can be identified with a certain concept. Again, it might be possible for two persons to have the same kind of brain-state, and yet for these states to constitute different concepts. Identity of concepts depends only upon identity (or sufficient similarity) of causal role.

But what if one envisages a whole system of such physical states, but not in their ordinary physical context? Have we then got a system of beliefs, concepts and other mental states? To this question I respond with another question. Suppose that we have an *isolated* DNA molecule, or suitable portion of such a molecule (or else something which, embedded in a cell, would have the same structure of causal powers). Have we then got genes or not? Perhaps the answer is a little arbitrary. But I favour saying that in such a situation we do have genes. So, equally, I incline to say that the isolated system of brain-states could still be said to be a system of mental states. No doubt it is a case of chess without the queen. But I would still call it 'chess'.

Incidentally, it has occurred to me since writing my original essay that genes (and so the DNA molecule) provide a good preliminary model for intentionality. On pages 150–1 I suggested that dispositional properties provide a first crude approximation to intentionality. The brittleness of a particular piece of glass 'points' towards breaking, although the breaking need never occur. Now consider the structure which, in a particular token of DNA, constitutes the blue-eyed gene. Its power to produce blue eyes in offspring is a dispositional property of the genetic material. But it is an especially sophisticated disposition. The genetic material is *encoded* in a very complex manner. This encoding will, *if all goes well*, produce blue eyes in offspring (in co-operation with much else besides, including other encodings). Now consider, say, the intention to paint the bathroom. I suggest that it stands to the actual painting of the bathroom somewhat as the genetic encoding for blue eyes stands to the blue eyes of the offspring.

Presumably it is an *objective fact* that a certain gene is 'encoded for blue eyes'. Equally, it is an objective fact that the intention to paint the bathroom is a state of some substance (physical, or non-physical) encoded for painting the bathroom. Analysis of even the simpler, biological, notion is not easy. But an analysis might pay great dividends. It might cast light on the more difficult notion of intentionality.

Theoretical entities. Malcolm finds my view that mental states are semi-observational, semi-theoretical, entities a curious one. If I intend to mow the lawn, and am aware of it (as one normally is of such an intention), then Malcolm finds it extraordinary to say that I am aware of it as 'a state within me impelling me towards mowing the lawn'.

Here I am reminded of a witty remark by Dr F. Knopfelmacher, of Melbourne University. He said of linguistic philosophy that it is 'ordinary language spoken in a plonking tone of voice'. Malcolm puts down my unordinary language, just because, if it were used in an ordinary context, it would be likely to mislead. There is, of course, appropriateness, subtlety and wisdom embodied in our ordinary ways of expressing ourselves. I agree with Malcolm that for me to *say*, when queried about my intentions for the afternoon, that 'there is a state within me impelling me to mow the lawn' sounds like the description of a neurotic impulse. All the same, I think that intentions are such impelling causes.

Why does making this remark in the course of ordinary chat sound so inappropriate? One thing involved seems to be that the form of words used does not differentiate between an intention, on the one hand, and, on the other, a mere *desire* to mow the lawn which, perhaps, I am not going to be swayed by. The description has not captured part of the special nature of *intentions*. A second, still more important, point is that such language *distances* one from the causal state in a way which, Malcolm correctly remarks, is characteristic of the way that we speak about neurotic impulses.

Here is an analogy. *Case 1*. An Australian says 'The Australians are going to fight'. *Case 2*. An Australian says 'We [he means the Australians] are going to fight'. The first remark is compatible with, and could conveniently be used by, a speaker who wanted to distance himself from the situation. He could go on to say 'But *I* am not going to'. The second remark could conveniently be used by a speaker who was identifying himself with his compatriots' purposes. It would be a bit surprising if he added 'But *I* am not going to'.

In the same way, I think, to say 'I am going to mow the lawn', *identifies* me with that causal factor within me whose presence I am asserting. I believe that it is the factor which will rule my afternoon. But I am also identifying myself with it, giving it official certification as a prospective act of D. M. A. I am speaking from the unity of myself, signalling that I am so doing. There is, if you like, something performative, and so not fully descriptive, in my use of language.

Malcolm goes on to say that one cannot observe one's own intention, and that an intention to do X does not cause one to do X. What he says seems to be simply a restatement of his own position. The only concession which I am inclined to make is that the language of announcing intentions does seem to have a performative, that is, a non-descriptive, element.

Malcolm also argues that if one seriously believes that intentions are brain-states, and one is asked what one is going to do this afternoon, it would be reasonable to reply 'How should I know? Let us wait to see what I will do.' But this argument begs the question against my position. After all, I contend that when I am aware of my own intention, although it is in fact a brain-state I am only introspectively aware of it as an (endorsed) factor impelling me in a certain direction.

Genuine duration. Malcolm reports me as arguing that if an intention does not lapse, then it continues in existence and therefore has genuine duration. The intention really is there in the mind all the time, and so can act as a cause.

I do not think that Malcolm has grasped the force of my argument here. Suppose that I intend, when next I see a certain person, to ask that person for a certain piece of information. We would all wish to distinguish the following cases. (1) Before I see the person, the intention lapses. (2) I see the person, the intention has not lapsed, but I forget to ask the question. (3) I see the person, the intention has not lapsed, and, *as a result*, I ask the question.

Now, if we take even a halfway scientific view of the matter (and that a scientific view of the matter is true, is a premiss of my argument), then we will think that certain causal factors must be present. In (1) it is natural to assume that a state, presumably a brain-state, corresponds to the intention, but that the state passes away before I see the person. In (2) the state continues, and is present when I see the person. But, because of inhibiting causal factors at work, it does not have the effect of my asking the question. In (3) the state continues, and, when I see the person, the latter perception plus the state (plus many other causal factors) brings it about that I ask the question. (A strange case, (3′), is possible where I ask the question, but not as a causal result of the state.)

If all this is denied, then I think that we are in the realm of magic and mystery. Human beings would not be within the causal order. Yet they manifestly are within that order.

But now, I argue, once it is conceded that such states exist, and have the causal powers which they have, then it is entirely natural (because intellectually economical) to go on to identify these states with the intention. These states have genuine duration, so intentions have real duration.

Of course, Malcolm thinks that he has good *a priori* arguments against the identification. Fair enough, but I claim to have met the arguments. In any case, the naturalness of the identification is itself some argument for thinking that Malcolm's arguments may be unsound.

Malcolm does make an interesting point when he remarks on the abstractness of the content of an intention. 'There is nothing to observe, to focus one's attention on, to monitor', he says. It has often been pointed out that intentions, and other conative states, yield little of the introspective detail associated with perceptual experiences and sensations. I think that I can suggest a biological explanation. Conative processes, by definition, set goals for the organism. Perception gives us the information by means of which these goals can be achieved. In a changing and uncertain environment goals need to be relatively abstract and flexible. But to achieve even these goals, information must be concrete and highly specific. It can be expected, therefore, that conative states will be relatively unstructured by comparison with perceptual states, and that introspective scanning of these two different sorts of state will involve some awareness of this difference.

Qualities of experience. The perceived colour, shape, taste, hotness or coldness, hardness and softness, etc., of a physical object appear at least to be intrinsic properties of the object. From these properties we gain the notion of a *sensible quality*. If, as I do, one thinks of introspection as like perception, but directed within to the mental, it will seem intelligible that introspection should equally make us aware of certain qualities. For instance, corresponding to perceived roundness or blueness, *perceivings* of roundness and blueness might be introspected as each having their specific qualities.

My belief is that there are no such introspected qualities, but I think it is an intelligible hypothesis that they exist. Malcolm draws attention to the way that, in ordinary language, we speak of various qualities of our sensations, calling a pain 'piercing', or 'dull', or 'gnawing', or 'throbbing'. But I doubt if his point is relevant here. I would take such characterizations of pain as indicating, in a rather rough but idiosyncratic way, something of the nature of the disturbance which feels to be taking place at 'the place of the pain'.

If we return to the topic of introspected qualities, Malcolm rejects the idea of specifying such qualities by means of an inner ostensive definition, or inward pointing. He says that such a procedure cannot fix the reference of any term. He may be right given certain conceptions of introspection, but they are conceptions of introspection which I reject. It is to be noticed that, on my view of introspection, our 'privileged access' to our own mental states is a purely empirical cognitive privileged access. (It is conceivable that A might have such access to B's mind, but lack it to his own, A's; while B have that sort of access to A's mind, but lack it to his own, B's.) Under these circumstances, I do not see why inner ostensive definitions should not be possible. The lack of intersubjective access would cut out direct ostensive teaching of one person by another, because of uncertainty about just what qualities were perceived. But inner ostensive definition would seem to be possible in principle.

I suggest, then, that the real question to be decided here between Malcolm and myself is whether my account of introspection is correct or not.

Causality. Malcolm compares two sentences: 'the strange move [of his oppenent at chess] caused him to be confused' and 'The drug caused him to be confused'. He says that what we have here are two different concepts of cause.

I wish to deny this. In both cases, I think, we have a chain of efficient causes. The causes and effects are *described* in somewhat different ways, but they are still causes in the same sense. In the first case, the opponent's strange move causes (*via* perception, inference, etc.) the player to believe that his opponent has made a strange move, a belief which causes him to become confused about what his opponent is up to. In the second case, the drug causes *general* mental confusion.

I do not think that Malcolm is right to say about the second case that the person is not confused or bewildered *at* anything. Rather, he is generally confused, that is, confused about a whole lot of things, perhaps confused in all his thoughts. He need not be confused about the drug, for he may not even know that it has been administered. In the first case, however, his confusion does have a specific object, the very thing which causes his confusion. Now I concede that if his state of confusion logically could be described only as a state of confusion at the strange move, then, unlike the drug case, it would be *essentially* internally connected with its cause. Such a difference in the two cases might well lead us to speak of different senses of the word 'cause'. But, of course, as Malcolm is aware, I hold that the state of confusion can be characterized independently of a characterization of its cause, *viz.* as a certain (presumably very complex) brain-state.

It is worth noticing in the drug case that we can set up an 'internal connection' between cause and effect by describing the confused state as 'the drug-induced state'. Compare the way that the sun can cause sunburn. I would say that in all cases of cause and effect the causal connection may be presented externally or internally, depending upon our choice of descriptive vocabulary.

Malcolm responds to this sort of reply by attacking the view that the confusion at the strange move could be a brain-state. He appeals to Miss Anscombe's argument again. What if just these states are produced in the brain of a person who knows nothing of chess, but is sitting in front of a chess-board with a player opposite him?

Malcolm's point here may depend upon a type–type identification of mental states with states of the brain, a sort of identification which, as I have already indicated, I do not accept. Again, if these states are not set in the right brain-context, and are not given the right causal powers, then they may not be beliefs about chess. Suppose, however, that Malcolm allows that the man's brain has been transformed in whatever way is necessary for these further conditions to be satisfied. Then, of course, the man has been miraculously turned into a chess-player.

Index